Pictorial Encyclopedia of FORESTS

Pictorial Encyclopedia of FORESTS

by Jan Jeník

Hamlyn
London · New York · Sydney · Toronto

PICTURE ACKNOWLEDGEMENTS

Aquila Photographics, H. Atanasová, ing. L. Bartoš, L. Brchel, Z. Bureš, ing. J. Cibulka, dr. V. Černý, S. Dalton, Natural History Photographic Agency, dr. M. Daniel, dr. L. Dobroruka, dr. D. Dykyjová, The Forestry Comission, ing. M. Grünerová, dr. J. Felix, J. Haager, dr. O. Hauck, L. Havel, B. Heger, ing. A. Horáček, dr. I. Hrdý, ing. M. Chroust, ing. J. Jeník, dr. J. Klán, dr. F. Kotlaba, dr. J. Kříž, dr. S. Kučera, L. Kunc, V. Lacina, dr. V. Liebl, dr. J. Lukavský, J. Marco, dr. P. Mazal, ing. I. Míchal, P. Pavlík, dr. ing. J. Pelíšek, I. Pekárková, dr. ing. J. Ponert, dr. J. Pradáč, dr. V. Prouza, ing. E. Průša, ing. K. Rataj, J. Rys, A. Říha, ing. K. Spitzer, dr. M. Spurný, dr. T. Sýkora, dr. K. Šťastný, ing A. Šika, Z. Thoma, dr. J. Vaněk, dr. Z. Veselovský, dr. V. Větvička, dr. P. Vlasák, J. Vogeltanz, dr. J. Volf, ing. M. L. Zelený, J. Zumr.

Translated by Olga Kuthanová
Designed and produced by Artia for
The Hamlyn Publishing Group Limited
Astronaut House, Feltham, Middlesex,
England

ISBN 0 600 33623 9
Printed in Czechoslovakia
3/05/08/51-01

CONTENTS

FOREWORD

Forests and woodlands are found throughout the world but the term 'forest' means something different in different countries and to different people. For an Icelander it means birch groves; for the European city-dweller it may mean oaks or beechwoods or the scent of pines; the timber merchant sees a forest only in terms of profit; the dweller in the tropics sees it as an impenetrable jungle; for the pygmy of Africa the forest is the whole universe, which begins outside the walls of his dwelling; the naturalist views it as a community of plants and animals; the geographer as only one of the earth's many formations.

There are many types of forest on the earth. Each is a complex community of numerous plant and animal species where many different physical and chemical processes are continually taking place. Is it at all possible to describe the complex nature of the forest in a simple way?

I always wanted to help people see more in a forest than what they had learned from experience in their immediate vicinity or what had been dictated by their specific interests. I still do not know how best to give the uninitiated an awareness of the broader workings of nature and of the many different interrelationships which surround them as they walk through the forest. I do not even know how to present in a single comprehensive volume the great diversity of the world's forests in a suitable form. I have tried, however, to do so here with the help of photographs.

A camera in the hands of a patient and trained photographer can capture in black and white or colour many of the basic features of the forest. I am indebted to many of my friends and colleagues who kindly contributed their pictures, which, together with mine, made it possible for me to write this book, intended as a guide for the reader in his contemplation of the illustrations.

One thing the reader should remember, however, is that, unlike the pictures and printed text, the realm of the forest is always four-dimensional — occurring in time as well as in space. The forest grows, multiplies, moves, smells, murmurs and sings. The forest is an intricate web of life.

Jan Jeník

1

1 An Alaskan landscape about 67° N. The meandering river and permafrost prevent total afforestation by White Spruce *(Picea glauca)*.

THE EARTH'S GREEN CLOAK

The sources of life-giving energy, water and food are distributed unequally over the earth's surface. This is due to the general arrangement of the land masses and oceans of the world and also to the movement of the planet relative to the sun. Throughout its yearly journey round the sun, the axis about which the earth itself rotates is tilted, and this gives rise to marked seasonal variations in temperature and varying lengths of day and night. In addition the sun provides the energy for the movement of air masses and the uptake of atmospheric water vapour. Wind force and direction as well as the humidity of the atmosphere is then further influenced by the distribution of seas and continents. These complex interrelationships have resulted in the formation of well-marked geographical regions which were determined long before plant and animal life appeared on the earth. During the various geological eras the earth's poles as well as the geographic zones shifted, but the distinct differences between the polar, temperate and equatorial regions remained.

Within the wide range of heat and cold, light and darkness, damp and drought, rich and poor soil, various forms of plant and animal life developed. Gradually extensive communities of plants and animals adapted to particular conditions of climate and environment evolved, and gave rise to such areas as tundras, forests, savannas, scrub lands, steppes, wetlands, mangrove woodlands and semi-deserts. Such associations of plant and animal life are called biomes. By their spread, diversity and beauty, forests, which cover nearly two-fifths of the earth's surface, are perhaps the most remarkable of the terrestrial biomes.

Prior to the spread of the human race forests covered practically the whole of Europe, and even to this day about a third is forested. (Fig. 2, 3, 4, 5 and 6). Most heavily forested is Finland, where trees cover about seventy per cent of the land and at the bottom of the list is Great Britain, with only about six per cent. In Asia vast forest complexes survive in the Siberian lowlands and on mountain slopes (Fig. 8) and there are extensive areas in the subtropical and tropical regions of south-east Asia. About two-thirds of both North and South America are tree covered. There are coniferous forests (taiga) in the far north (Fig. 1), mixed forests of evergreens and deciduous trees (which shed their leaves annually) and palm trees in the subtropics (Fig. 7), and evergreen rainforests in the Torrid Zone (Fig. 9). Forests cover a quarter of Africa and about a fifth of Australia (Fig. 10).

Antarctica has no forests now though they did exist in the distant geologic past. The cold climate is also the reason for the absence of forests in the far north and high up in the mountains. Forests, however, are to be found even where summer is brief and lasts for only a few weeks, and where winter is long and cruel (Fig. 11 and 12). Trees line the slopes wherever they are not thwarted by gales, snowdrifts and avalanches (Fig 13 and 14). In the arctic tundra and high mountains, pioneer trees, the vanguard of the forest, are mostly spruces *(Picea)*, firs *(Abies)*, larches *(Larix)* and pines *(Pinus)*. On Mount Olympus in Greece, for instance, the tree growing highest up is the Bosnian Pine *(Pinus leucodermis)* (Fig. 18) while in the Himalayas it is the Indian Silver Fir *(Abies spectabilis)* and various species of spruce, hemlock *(Tsuga)*, yew *(Taxus)*, and juniper *(Juniperus)* (Fig. 15). Even in the high mountains of the tropics the forest must come

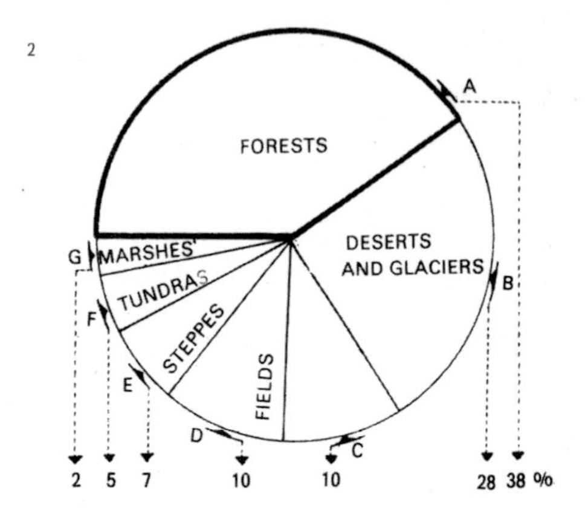

2 Percentage of land area which could potentially be occupied by forest compared with other terrestrial biomes.

3

4

3 Vast stands of Holm Oak *(Quercus ilex)* in the Pyrenean foothills.

4 Mixed forests with the European Silver Fir *(Abies alba)*, the Common Beech *(Fagus sylvatica)* and the Norway Spruce *(Picea abies)* in the West Carpathians.

5 A dark mantle of evergreen Norway Spruce *(Picea abies)* covers Europe's mountains.

5

6

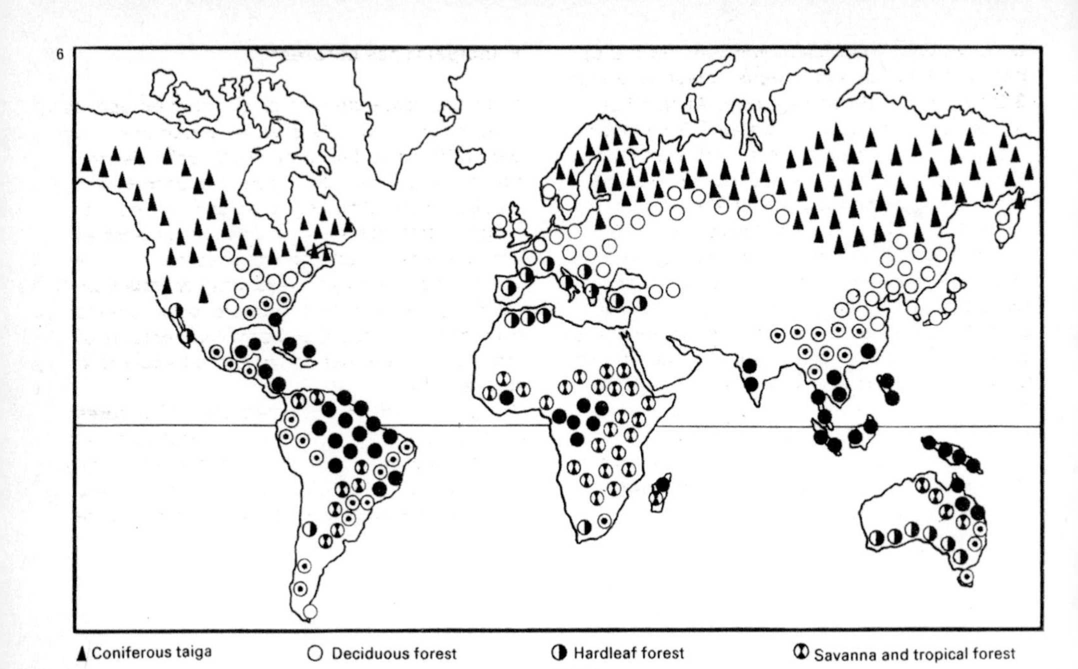

7

6 Distribution of the main types of forest throughout the world.

7 Stands of Royal Palm *(Roystonea regia)* in the Sierra del Rosario Mountains in Cuba.

8 In the Himalayas fir and spruce forests are found as high as 3,500 metres above sea level.

to terms with insufficient warmth and the uppermost limit of continuous forest is about 3,000 metres above sea level. At this height on Africa's highest mountain, Kilimanjaro, there are forests composed of the softwood *Podocarpus milanjianus* and hardwoods such as *Hagenia abyssinica, Ilex mitis, Hypericum revolutum* and *Ocotea usambarensis.*

The spread of forests is also kept in check by shortage of water, by fire and by poor soil or wetlands with insufficient oxygen for root growth. However, there are always pioneers which can gain a foothold even in such harsh conditions. In such places forests are usually thin and the trees small and crooked or scrub like. The roots and leaves, as well as the reproductive organs, are adapted to overcome the inhospitable conditions of the environment (Fig. 16, 19, 20, 21 and 22).

Principal types of forest

With the vast extent of its distribution and its adaptation to widely diverse conditions, it is very difficult to find a generally valid description for the forest. In different countries it includes areas differing markedly in density, height, stratification and diversity of species. For example, many forest preserves in the subtropics would be looked upon as little more than a park or a fruit orchard by a forester from central Europe, and park-like forest is often found where dense forest meets tundra or steppe.

Typical distinguishing features of a forest are the height of the crowns above the ground and the depth of the roots in the soil. Also important are the numbers of stems on a given area, the density of crown cover and the

8

9 A mantle of tropical forest cloaks the mountainous landscape round Machu Picchu in Peru.

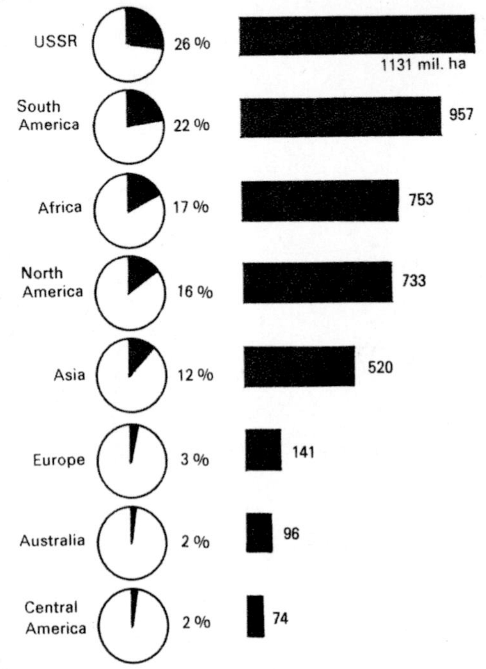

10 Area covered by forest in various parts of the world: right — total area in millions of hectares; left — percentage of total forest area in the world.

total area covered by a stand of trees. Generally classed as a forest is a biome with trees taller than five metres on an area of at least a hundred square metres, with crowns covering a minimum of one-third of this area. Where the dominant trees are a great deal taller than five metres and the forest is very open then it is said to be a tree savanna or savanna woodland. The determining factor which establishes a forest as such is whether the trees already form a distinctive environment. A thick avenue of tall trees could hardly be looked upon as a forest (Fig. 25).

Forests are often divided into three groups: coniferous, broadleaved (deciduous) and mixed. This practical division apparently originated in Europe and North America where the term coniferous trees is synonymous with gymnospermous trees which are generally cone bearing and usually have narrow leaves or 'needles' and broadleaved trees which is synonymous with angiospermous dicotyledonous trees, which are flowering trees with net-veined leaves. Coniferous trees have many peculiarities in anatomical structure and physiology, including soft wood which can be easily processed. The wood of broadleaved trees,

In early times much of Scotland and parts of northern England were covered with Scots pine forest. Remnants of these ancient forests still exist, the old trees growing in a carpet of heather.

Autumn in the beech forest.

11 A pioneer island of spruce forest amid permafrost on the Yukon River, Alaska.

12 In the Alaska Range forest is limited in height along the polar timberline.

11

on the other hand, is usually hard and more durable, and these trees are commonly called hardwoods. In the tropics, however, the division into coniferous, broadleaved and mixed forests is less practical. For example many species of the gymnospermous genus *Podocarpus* and the widely distributed Kauri *(Agathis australis)* have broad to ovate leaves, while certain dicotyledonous trees of the heather family (Ericaceae) have narrow linear leaves. Furthermore, a very distinctive element in tropical forests is palm trees, which are angiospermous, monocotyledonous trees—having parallel-veined leaves.

Principal types of trees and their classification

Phylum	Lower unit	Example—genus	Predominant shape of leaf	Persistence of leaves
Pteridophytes	ferns	*Cyathea*	crown of pinnate leaves	evergreen
Gymnosperms	cycads ginkgo conifers	*Cycas* *Ginkgo* *Pinus*	crown of pinnate leaves broadleaved needle-like, occasionally broadleaved	evergreen deciduous evergreen occasionally deciduous
Angiosperms	dicotyledons monocotyledons	*Fagus* *Raphia*	broadleaved, occasionally needle-like crown of pinnate or fan-shaped leaves	evergreen and deciduous evergreen

12

13

The difficulties are even greater when it comes to distinguishing between deciduous and non-deciduous forests. This division stresses the fact that some trees shed their leaves annually and are bare part of the year whereas others are permanently covered with leaves or needles, which are never shed all at one time. In actual fact, however, all trees and shrubs shed their leaves, which are generally short-lived organs, capable of functioning either for only a few months, a single year or at most for ten years. In the temperate regions of Eurasia and North America most broad-leaved trees and larches (which although they are conifers are deciduous) shed their leaves in the autumn and produce new ones the following spring, whereas most conifers retain their needles even in winter so that they appear to be evergreen. The length of time a needle remains on the tree depends upon its height above the ground, the amount of light it receives and the action of the wind. The Common or Norway Spruce *(Picea abies)* retains its needles for six to nine years (and occasionally for as long as twelve years) while the European Silver Fir *(Abies alba)* keeps them for about ten years on average.

The needles of most pines have a shorter life span. In the case of the Scots Pine *(Pinus sylvestris)*, for example, a pair of needles is shed after about three years. Pines shed not only the needle-like leaves but also the short twigs (brachyblasts) from which the needles grow in twos, threes, fours, fives and very occasionally also singly, as in the Singleleaf Pinyon *(Pinus monophylla)* of the high mountains of California, Nevada and Arizona.

The matter of shedding leaves is even more complex in the case of monocotyledonous trees. Forest palms always shed the oldest marginal leaves in the crown. A palm leaf remains attached for several months or even for several years, depending on the species, the location and the rate of growth. In view of the large size of many palm leaves the period when they are shed is a dangerous time to enter a palm grove, for a dry leaf with its heavy stalk generally weighs several kilograms.

Evergreen crowns are not restricted to conifers and palms; many dicotyledonous trees have them too. Evergreen broadleaved trees are common in regions with a Mediterranean climate such as southern Europe, North and South Africa, California and Indochina. Typical, for instance, are evergreen oaks or members of the laurel family (Lauraceae). Most evergreen broadleaved species are to be found in damp tropical forests where the temperature and humidity remain more or less constant throughout the year. Evergreen broadleaved trees, however, are to be found also in countries with marked periods of drought or cold. New Zealand has extensive forests made up of various species of southern beeches *Nothofagus* that keep all their foliage the whole year long.

Forests with a single dominant species of conifer or broadleaved tree may be divided into the following groups:

(1) deciduous coniferous forests (larch forests

in Siberia, Swamp Cypress *(Taxodium distichum)* forests in Florida)
(2) evergreen coniferous forests (spruce, fir, hemlock or pine forests in Europe and North America)
(3) deciduous, broadleaved forests (the oak and beech forests of Eurasia)
(4) evergreen broadleaved forests (the laurel forests of Mediterranean regions and Nothofagus forests of New Zealand)

Many forests, however, do not have a single dominant species but are composed of several or even many species of either broadleaved trees or conifers, or even a mixture of both. The various species may differ as regards the persistence of the leaves. Some may shed their leaves at a certain time, others at a different time, and some may not shed their leaves at all (more precisely they do not shed their leaves all at once at a given time). Such unclassified mixtures may be found, for instance, in the vast spaces of Siberia and Canada. They include mixtures of deciduous larches and non-deciduous pines and spruces, as well as mixtures of both types of conifers and deciduous birches, poplars and alders. Much more complex are the mixtures of deciduous and non-deciduous forms in tropical regions, where periods of damp alternate with periods of drought. The general term used to describe these forests is semideciduous, but the word has a number of different meanings. Some such forests may be partly bare during the dry season because they contain species that regularly shed their leaves at this time; others are partly bare because various species shed and grow new leaves at different times of the year; still others are semideciduous because under the given conditions, various trees of the same species shed their leaves at different times. The environment in tropical forests did not force trees to adopt the same life rhythm — either as populations, individual trees, or even within the crown of a single tree.

The appearance of coniferous forests varies with the climatic and soil conditions as well as

13 Snow, frost and wind prevent the spread of forest in the mountains.

14 European Larches *(Larix decidua)* are the forest pioneers at high elevations in the Alps.

15 Coniferous forest and rhododendrons growing at above 3,000 metres in the Himalayas.

14

15

16

17

16 In areas with unstable moisture conditions or in places that are completely arid, the forest is equipped with life forms that tolerate lack of water. Savanna woodland with the palm *Colpothrinax wrightii* on the Isle of Pines in the Caribbean.

18

19

20

with the structural characteristics of the dominant species. In the far north and in mountains (Fig. 29 and 30) spruce forests usually have a thin canopy and branches growing all the way up from the ground. Light-loving pines, on the other hand (Fig. 27), usually have a smaller crown in old age and a bare bole without branches. In good soil and in a congenial climate, coniferous forests consist of tall trunks of great thickness.

Broadleaved forests differ in character in various parts of the world. The shape of the trunk and crown is adapted to the conditions of the environment. In deep and damp soil the trees have tall, slender boles and a crown placed high up on the trunk (Fig. 28 and 31), while in shallow or dry soil the same species may make only dwarfed and twisted trunks (Fig. 33). Unlike the broadleaved trees of Europe and America, the dicotyledonous trees of the equatorial virgin forests exhibit much greater diversity in the shape of the trunk, crown, leaves and roots (Fig. 32, 34 and 35). Sclerophyllous forests, made up of trees with stiff, leathery leaves (olives for example), are a group apart; they are common in the median latitudes of both hemispheres, which have a Mediterranean climate.

Mixed forests composed of conifers and broadleaved trees are commonly found on the dividing line between coniferous and broadleaved forests. In Europe, there are beech-fir forests (Fig. 36 and 37) composed of the Common Beech *(Fagus sylvatica)* and European Silver Fir while in North America there are mixtures of maples *(Acer)*, oaks *(Quercus)*, birches *(Betula)*, pines and hemlock.

21

17 In the high mountains of Africa mist forest grows up to altitudes of 3,000 metres.

18 The hardy Bosnian Pine *(Pinus leucodermis)* stands firm on the slopes of Mt. Olympus in Greece.

19 Only drought and fire prevent the spread of tree savanna in West Africa.

20 The Common Beach *(Fagus sylvatica)* and Sessile or Durmast Oak *(Quercus petraea)* grow even on stony soil.

21 Nature solved the problem of permanently waterlogged soil by developing specialized forests. *Aeschynomene elaphroxylon* in Africa.

22

22 The forest has adapted successfully to deep peat. Stands of the pine *Pinus rotundata* in Czechoslovakia.

Main Types of Forest

Distinguishing characteristics	Name of Forest	Occurrence
Type of leaves	Coniferous forest, taiga Broadleaved forest Sclerophyllous or laurel forest	Canada, Siberia Central Europe Mediterranean region
Type of tree	Deciduous, broadleaved forest Deciduous, coniferous forest Semideciduous forest Evergreen, broadleaved forest Evergreen, coniferous forest	Central Europe Eastern Siberia (larch) South-east Asia Mediterranean region, the tropics Canada, Siberia
Density of growth	Closed forest Open forest Savanna woodland Forest-steppe	Dense and tall forest anywhere Lower, unstratified forest Open woodland with grass Landscape with alternating expanses of forest and steppe
Source of water	Rainforest Mist forest Swamp forest Peat forest Floodplain forest Gallery forest Mangrove woodland	In regions with great precipitation In high mountains alongside rivers In swamps In peat bogs In river valley plains alongside rivers Alongside rivers in dry country On the seacoast in the tropics
Altitude	Lowland forest Mountain forest Subalpine forest	On flat land and in foothills On mountain slopes Below the timber-line in high mountains

The composition of forests is greatly influenced by their geographic location, or more precisely by their latitude, longitude, altitude, and distance from the sea, and by the prevailing currents and winds. Geographic location determines the basic supply of heat and water and accessibility to colonizing plants and animals. The climate of the various geographic zones is the basic criterion for the division of forests.

(1) The far northern regions of Eurasia and North America are covered with coniferous forests (taiga). Similar forests (mountain taiga) may be found on all high mountains.

(2) In the temperate regions of Eurasia and North America the forests are composed of deciduous broadleaved trees, giving way to mixed forests towards the taiga.

23 Structure of a tropical evergreen forest.

24 Structure of a tropical semideciduous forest.

23

24

25

27

25 A dense double avenue is not a forest for it does not form an environment for typical forest plants and animals.

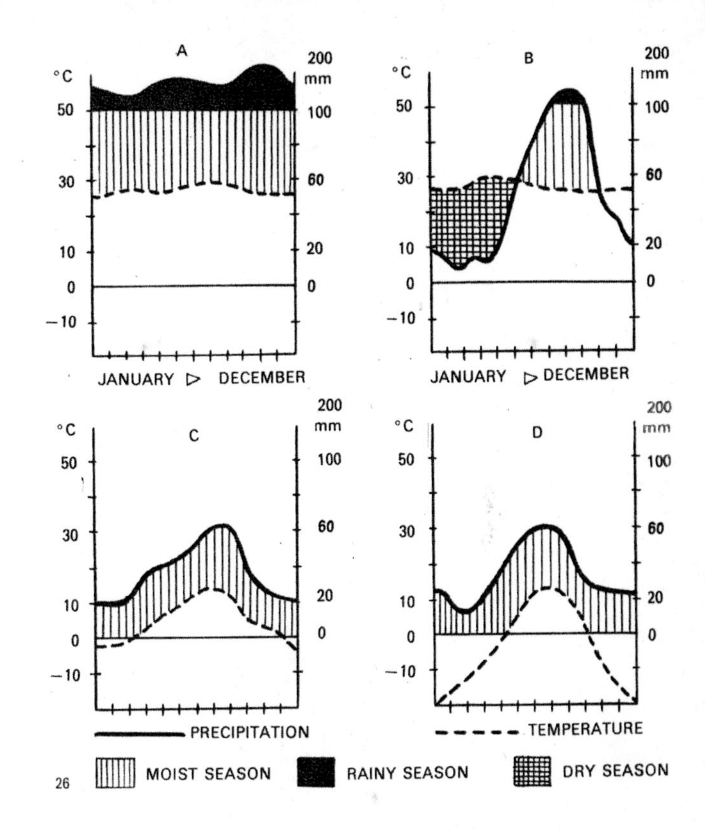

26 Features of the climate of four forest regions according to the mean monthly temperatures and mean monthly precipitation; top left — tropical forest; top right — savanna woodland; bottom left — temperate deciduous forest; bottom right — northern coniferous forest.

27 Light-loving pines generally form open forests like this one in southern France.

(3) In regions of both hemispheres having a Mediterranean climate the forests are non-deciduous, sclerophyllous (southern Europe, north Africa, California, Chile, southern Africa, Indochina and Australia).
(4) In the subtropical and tropical regions where the year is divided into a rainy season and dry season there are continuous savanna woodlands and open tree savannas.
(5) The tropical and equatorial regions with constant precipitation throughout the year are covered with dense, tropical, evergreen forests (rainforest).

The dominant species determine the character of the world's forests and are always the decisive element in the life of the entire biome, though important roles are also played by all the other plants and animals that live there. The presence of specific forest orga-

28

29

28 Oaks *(Quercus)* are the dominant trees of the deciduous broadleaved forests of Europe, Asia and America.

29 Slender, columnar trees are typical of the coniferous forests round the polar limit.

30 The typical structure of a coniferous virgin forest is shown by this forest of Norway Spruce.

31 Common Beeches *(Fagus sylvatica)* form typical deciduous forests.

30

nisms in a certain region is the result of the lengthy evolution of life on the earth. In previous geological epochs the continents were not as they are now, and oceans formed other and different barriers to the spread of plants and animals. The climate also changed with the passage of time as did the soil composition and water balance. The most radical changes, however, occurred in life itself.

The evolution of forest

Three hundred million years ago during the Carboniferous and Permian Periods of the Palaeozoic era, great coal swamp forests flourished on the earth. In the damp tropical climate they laid down layers of decomposing matter in the swampy soil which over the aeons, turned into coal. These forests contained tree-like ferns, horsetails, club mosses and cycads. The following geological era—the Mesozoic — saw the gradual decline of horsetails and the appearance of the first conifers, among them the Maidenhair Tree *(Ginkgo biloba).* Fossil finds prove that 150 million years ago forests of these trees covered large parts of Eurasia and North America. Amazingly enough this species has survived to the present day, growing in small stands in south-eastern China. Other trees that have been on the earth for a remarkably long time are the sequoias *(Sequoia sempervirens, Sequoiadendron giganteum)* and the Bald Cypress *(Taxodium distichum)* which form large natural forests in North America. Great excitement reigned amongst botanists thirty years ago when a living *Metasequoia glyptostroboides,* previously known only as a fossil, was discovered in China.

In the Tertiary period of the Cainozoic Era a great leap forward was made with the appearance of the first flowering plants (angiosperms), which included trees growing to this day, such as magnolias *(Magnolia),* plane trees *(Platanus),* tulip trees *(Liriodendron),* breadfruits *(Artocarpus)* and figs *(Ficus).* These were followed in time by oaks, beeches, walnut trees *(Juglans),* maples, birches, willows *(Salix)* and poplars *(Populus).* There was also an increasing number of various monocotyledonous trees such as palms, bamboos and pandanus. The surviving conifers and rapidly evolving dicotyledonous and monocotyledonous trees formed the basis of our modern forests.

In the late Tertiary the evolution of forests

31

32

32 The structure of tropical deciduous forests is distinguished by an exceptionally dense canopy and occupation of all the layers. Here tree ferns of the species *Cyathea manniana* grow on the forest floor.

33 On dry, steep banks deciduous forests are made up of only small, crooked trees.

33

was affected by the gradual cooling of the climate, which caused the extinction of species which could only thrive in comparatively hot conditions (thermophilous species). It was also affected by the ice ages when vast ice sheets covered the northern half of Europe and North America several times in succession. This resulted in the extinction of trees which once grew in Europe and which are still common in American forests, such as hickory *(Carya)*, tulip tree, sweetgum *(Liquidambar)*, and many conifers, including hemlocks, arbor vitae *(Thuja)* and redwoods *(Sequoia)*. In North America, forests had greater freedom of movement during the ice ages for the mountain chains there are positioned north to south and the continent extends as far as the tropics (Florida, the Caribbean region, Central America). When the climate became warm again as the ice sheet receded, forest plants and animals were able to follow in its wake and return north without any difficulty. This accounts for the forests of North America comprising a greater variety of species than is found in Europe.

Principal Periods in the Evolution of the Forest in Central Europe Following the Last Glaciation

1. Late Glacial Period (beginning between 15,000 and 20,000 years ago)
   - I. Lower Dryas Period; tundra vegetation
   - II. Alleröd Period: pine and birch forests
   - III. Upper Dryas Period: tundra vegetation
2. Post-Glacial Period or Holocene (beginning 10,000 years ago)
   - IV. Preboreal: pine and birch forests
   - V. Boreal: widespread distribution of hazel
   - VI. Early Atlantic: mixed oak forests
   - VII. Late Atlantic: mixed oak forests and appearance of beech-fir forests
   - VIII. Sub-boreal: mixed oak forests and beech-fir forests
   - IX. Sub-Atlantic: beech-fir and hornbeam forest
   - X. Sub-recent: cultivated coniferous forests and man-made steppe

In some parts of the world the evolution of forests in the later geological periods (in the Tertiary and Quaternary) was unimpeded by

34

34 Often found below the canopy of tropical rainforests are dwarf trees bearing a crown of leaves at the top of the trunk.

35 Savanna woodlands are low and have a broken canopy.

35

large-scale catastrophes. The forests of the tropical lowlands in particular were saved from glaciation and sudden changes in climate and this contributed to their great wealth of species. This is evident in south-east Asia, the Amazon basin and the Zaire River region, and it is for this reason that the forest flora of the Malaysian peninsula, for example, numbers 8,000 species of vascular plants—double that of Central Europe which is a much larger and more diversified territory.

The distribution of forests

The diversity of the earth's forest cover is perhaps best shown by three broad imaginary bands running vertically down the continents (Fig. 38). The first band extends from northern Europe south across central Europe, the Mediterranean, north Africa, the Sahara, and Zaire to Capetown in the south of Africa. The second runs from the north down eastern Siberia, China, the high mountains of central Asia, India and Southeast Asia to Australia. The third extends from the forest tundras of northern Canada through the forest regions of the eastern and western United States, on to the Caribbean, Central America and down the South American continent.

Band one

At the forest limit in the north of Europe grows the White or Downy Birch *(Betula pubescens)*. Its resilient trunk, flexible branches and deciduous leaves enable it to stand up well to the harsh arctic winters, lengthy snow cover and cold summers. Further south, in the forests of Scandinavia and northern Europe, the two prevailing trees are conifers, namely the Scots pine and the Norway spruce (Fig. 39 and 40, 42). The extremely vigorous spruce is unable to flourish in poor soil, on rocky slopes or in waterlogged soil and here its place is readily taken by the pine and pioneer birches, alders and willows. Pine also covers extensive tracts devastated by fire—the feared scourge of northern coniferous forests.

36

37

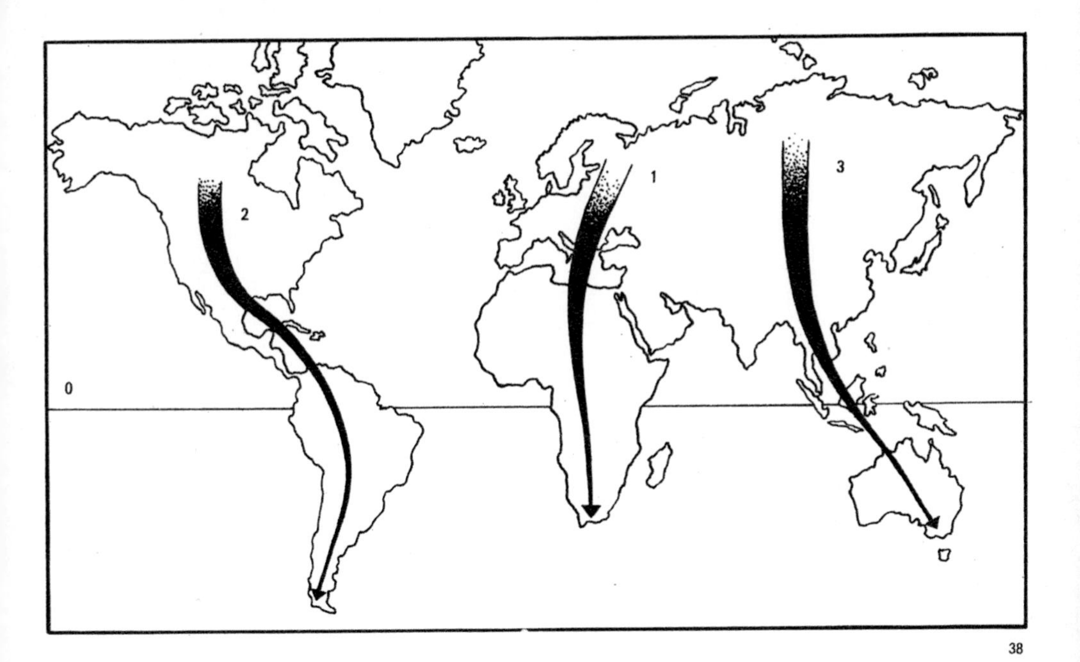

38

36 Mixed forests composed of Common Silver Fir *(Abies alba)*, Norway Spruce *(Picea abies)* and Common Beech *(Fagus sylvatica)*.

37 Mixed forest with Norway Spruce *(Picea abies)* and Common Beech *(Fagus sylvatica)*.

38 Diagram showing three imaginary bands running from north to south and pertaining to the description of the world's forests.

39 Deformed stands of Scots Pine *(Pinus sylvestris)* and Common Beech *(Fagus sylvatica)* on the coast of the Baltic Sea.

39

40

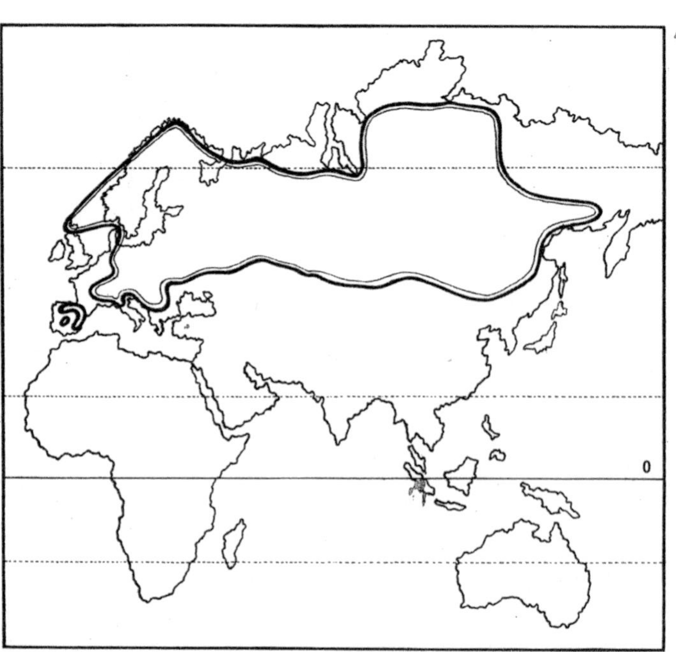

40 The taller mountain ranges and northern parts of Europe are dominated by forests of Norway Spruce *(Picea abies).*

41 Distribution of the Scots Pine *(Pinus sylvestris)* in Eurasia.

At the upper limit of a mist forest on Kilimanjaro, 3000 metres above sea level *(Hagenia abyssinica, Podocarpus milanjianus, Lobelia giberroa).*

A tropical forest containing eucalyptus trees and palms in the mountains on the eastern coast of Australia.

An evergreen forest of Indian Silver Fir *(Abies spectabilis)* and shrubby rhododendrons at an altitude of 3,900 metres in the Barun Khola Valley of the Himalayas.

The Scots Pine *(Pinus sylvestris)* is a remarkable tree, stands of which cover a vast territory extending from Spain, Great Britain and Scandinavia in the west, across Europe to the shores of the Sea of Okhotsk in eastern Siberia. In size of range it is unrivalled by any other pine, even though the Jack Pine *(Pinus banksiana)* of America and two American spruces *(Picea glauca* and *P. mariana)* also have a vast range reaching from Alaska to Newfoundland (Fig. 41).

The Scots pine is noted for its great adaptability to diverse climatic and soil conditions. It spread as a pioneer in the wake of the ice-sheet in northern and central Europe, colonizing these areas as far back as 15,000 years ago. Analyses of fossil pollen found in peat deposits reveal that the tundra in these areas had been replaced by Scots pine together with birch by 12,000 years ago. After this the pine was forced back by a new period of cold only to return again in the Preboreal as the dominant tree of northern and central Europe.

The Scots pine's major limitation is its small competitive vigour. More than anything else it requires light in order to grow well and if it is shaded by other trees then adult specimens show poor development and young seeds will not even germinate. In the postglacial period, therefore, it was gradually forced to yield the territory it had colonized to the more vigorous spruce, fir, beech and other trees. Only under particular conditions does it tolerate association with these trees, though it will form permanent mixed communities in areas having rich sandy soil. Despite their limitations pine forests found suitable conditions in a number of places, chiefly in Scandinavia, Siberia, in the lowland country of both East and West Germany and in Poland.

42

43

44

42 On the highest ridges spruces give way to dwarf Swiss Mountain Pine *(Pinus mugo)*.

43 Beech, fir and spruce are mixed to form the main type of forest at median elevations in European mountain ranges.

44 On the banks of mountain rivers, the forest begins with pioneer stands of willow and tamarisk.

The Scots pine has been saved from the spread of stronger trees, particularly the spruce, in these areas because all disruptive catastrophic factors contribute to its spread and establishment. In places swept by gales and particularly in places devastated by fire, the pine spreads rapidly thanks to its ever present and easily dispersed seed. Its chief rival, the Norway spruce, is completely destroyed by fire because its thin bark does not provide the living tissues within the trunk with sufficient protection from the damaging effects of heat. The pine, on the other hand, is protected by thick bark and furthermore, hardy specimens which are frequently found growing on inaccessible rocks usually survive such a holocaust. Their winged seeds, which find burnt soil without humus particularly conducive to germination, are then dispersed by the wind and afforest the area anew. Scots pine always produces great quantities of seed at intervals of three to five years and the seed has very good powers of germination, remaining viable

45 The valleys of large European rivers contain highly productive broadleaved forests with an extraordinary multiplicity of species.

46 Meandering rivers in Europe are bordered by broadleaved forests, either of natural origin or planted by man.

45

46

47

47 On alluvial soil the Common Alder *(Alnus glutinosa)* follows in the wake of reed beds.

for about three years. This, then, explains why in Scandinavia there are such large areas covered with pure stands of pine. It takes several hundred years for the vigorous spruce to oust the pine completely, and longer if it is halted by a new forest fire.

Scots pine is a diverse species comprising a number of geographical and ecological races, which at first glance are hard to identify according to common morphological characters, such as needles, bark or cones. Extensive comparative studies revealed that Scots pines of different provenance show marked differences in the life cycle of the individuals from germination to maturity.

Southward from northern Europe, on all richer soils in the foothills and mountains of the central part of the Continent are forests of

48 Thermophilous broadleaved forests by the Neretva River, Yugoslavia.

49 In southern Europe and in the Balkans the dominant trees are the Downy Oak *(Quercus pubescens)* and the Eastern Hornbeam *(Carpinus orientalis)*.

beech and European silver fir (Fig. 43). In warmer places on south-facing hillsides, in the warmer regions of southern Europe and in all poorer soils where beech does not thrive, the forests contain oak, European Hornbeam *(Carpinus betulus)* and pine. Particularly rich in species are forests which grow on valley bottoms and flood plains, where are found large numbers of ash *(Fraxinus),* alder, poplar and willow (Fig. 44, 45, 46 and 47). Spaces from which forest trees have been removed by death or some catastrophe generally contain birch, aspen *(Populus tremula)* and mountain ash *(Sorbus).* The warmest slopes and lowlands in the south of Europe (Fig. 49) are covered with various thermophilous trees such as the Downy Oak *(Quercus pubescens),* Turkey Oak *(Quercus cerris)* and the Hungarian Oak *(Quercus frainetto).* Still further south, in the lowlands and hill country of Macedonia and Greece, the dominant trees of the broadleaved forests are species with demanding heat requirements (Fig. 48, 50 and 51) such as Eastern Hornbeam *(Carpinus orientalis),* Manna Ash *(Fraxinus ornus)* and Holm Oak *(Quercus ilex).* On rocky and sandy soils there

50

50 Shrubby remnants of broadleaved forests in the limestone mountains of Yugoslavia.

51 In Albania remnants of broadleaved forests are to be found only at the foot of mountains and near rivers.

51

52 Mixed forest with Italian Cypress *(Cupressus sempervirens)* and Aleppo Pine *(Pinus halepensis)* on the Mediterranean coast of Yugoslavia.

is ample space for various species of pine, the Mediterranean Cypress *(Cupressus sempervirens)* and other undemanding trees (Fig. 52 and 53). There are also forests of Common Larch *(Larix decidua)* and the lovely Arolla Pine *(Pinus cembra)*, both of which have related forms in Siberia (Fig. 54).

In the lowlands of north Africa the only existing forests are located far from human habitations. They are the remains of once extensive forests of Holm Oak, Cork Oak *(Quercus suber)*, wild species of olive *(Olea oleaster)*, Carob *(Ceratonia siliqua)*, and Maritime tropical Pine *(Pinus pinaster)*. In the high Atlas

52

Mountains are to be found the last remnants of stands of Atlantic Cedar *(Cedrus atlantica)*, related to the Cedar of Lebanon *(Cedrus libani)*, whose fate in the mountains of Asia Minor is also already sealed.

Further south lies the Sahara, bordered only by tree or shrub savannas, where they have not been destroyed by man. Towards the equator the tree savannas give way to the savanna woodlands of the Guinea zone which contain trees of greater stature, such as *Isoberlinia doka* (Fig. 57), *Daniellia oliveri*, *Lophira lanceolata*, and various acacias *(Acacia)*.

53 Remnants of cedar forests *(Cedrus libani)* in the mountains of Lebanon.

54 Altitudinal forest belts in central Europe 1. mixed forest with oak dominant, 2. oak and hornbeam forest, 3. beech and fir-beech forest, 4. spruce forest, 5. dwarf pine.

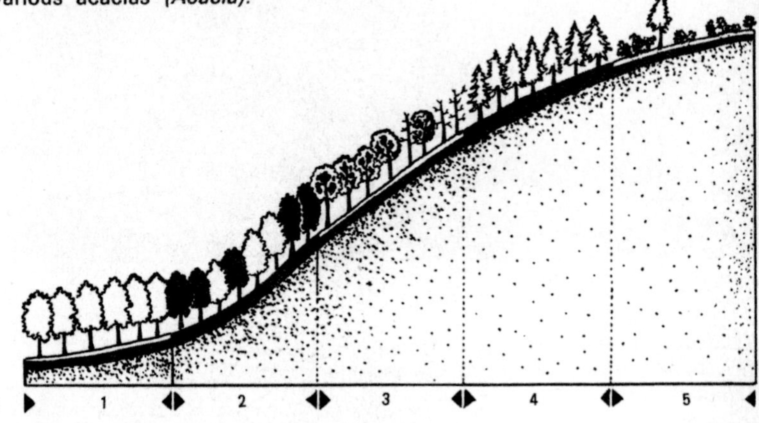

55

55 Stony banks at the edge of the rainforest in West Africa are covered with the endemic *Talbotiella gentii.*

Coast redwoods *(Sequoia sempervirens)* in the coastal region of California near San Francisco.

Mixed forest with maples *(Acer)*, oaks *(Quercus)*, birches *(Betula)* and Canadian Hemlock *(Tsuga canadensis)* on the eastern coast of Canada.

Coniferous forest with dominant Hartweg Pine *(Pinus hartwegi)* on Mount Popocatepetl.

Flood plain forest at a tributary of the Amazon River in Brazil.

Mountain forest at an altitude of 2,500 metres in the Venezuelan Andes with silver-leaved cecropias *(Cecropia telenitida, C. santanderensis)* and *Podocarpus* species.

This is also the home of the memorable Baobab *(Adansonia digitata).* Deciduous savanna woodlands are interspersed in places with dry evergreen forests (Fig. 55) which link up with the evergreen tropical rainforests around the equator (Fig. 56, 58 and 59). Africa's rainforests contain several thousand species of trees and are generally rich in flora and fauna. The upper tree layer of giants is composed mostly of members of the Caesalpiniaceae, Meliaceae, Sterculiaceae, Anacardiaceae, Moraceae, and other families.

In the southern hemisphere, in Angola and Zambia, large tracts are covered by dry tropical forests called miombo in which the dominant trees are members of the genus *Brachystegia* and *Julbernardia.* Still farther south conditions are not very propitious for the growth of forests. The velds of Kalahari and Botswana consist mostly of grass and thickets and there are trees only in the vicinity of larger water courses. Sclerophyllous evergreen and deciduous forests appear again only in a narrow strip along the eastern coast of South Africa and in the southernmost part of the country.

The whole continent of Africa shows with remarkable clarity how the longitudinal forest belts are disrupted by the influence of climatic anomalies, mountain topography and unusual soil substrata. South of the Sahara a belt of savanna woodlands extends from the western edge of the continent (from Senegal and Guinea) eastwards to the high mountains of Ethiopia, where, because of the altitude and the mixed pattern of regional climates, very different forests appear. The belt of equatorial rainforests is likewise not entirely continuous. It begins in the west in Sierra Leone and continues across Liberia and the Ivory Coast to southern Ghana. The same type of forest reappears in Nigeria, Cameroon, Gabon and Zaire. The rainforest belt is thus interrupted in southeast Ghana, Togo and Dahomey, where savanna woodlands extend from the north all the way to the coast of the Gulf of Guinea. This important hiatus, called the Dahomey Gap, marks the biogeographic limit in the distribution of many forest trees, herbs, animals and forest communities (Fig. 62).

For a long time the reason for this unusual break in the belt of rainforest along the coast of the Gulf of Guinea remained a mystery. New investigations reveal that it was caused by the Guinea ocean current which flows along the coast from the west past the jutting Cape Three Points and create off the coast of

56

57

56 The structure of Africa's rainforests is determined by the giant species of the mahogany (Meliaceae) and Caesalpiniaceae families.

57 *Isoberlinia doka* is the dominant species of tree in the savanna woodlands south of the Sahara.

58

58 Forests opened by farming are soon invaded by lianas.

59 Important to the structure of the tropical forests of America are the aerial roots of aroids (Araceae) which successively become anchored in the soil and become taut as their tissues contract.

59

Ghana and Togo huge whirlpools which bring up cold water from the depths. This cools the surface air, which moves towards and over the warm continent (particularly during the day) causing a rapid drop in atmospheric humidity and limiting the condensation of water vapour. The result is that the annual rainfall is only 700 millimetres and this is very irregularly distributed throughout the year. Such conditions make the growth of rainforest impossible and only savanna with scattered trees and short grass is found here.

In the southern parts of Africa, too, the transverse forest belts are disrupted by climatic anomalies caused by ocean currents. Along the whole south-west coast from Angola past Namibia to South Africa flows a very cold ocean current that creates steppe, semidesert and desert conditions in a wide belt bordering the coast. In Angola, savanna woodlands therefore begin far back in the interior, and the coast of Namibia is bleak desert. In contrast, the coastal belt on the other side of southern Africa between Laurenço Marques in Mozambique and Port Elizabeth in South Africa has markedly moister conditions than the interior and thus has a diversity of forests and savannas (besides the mangrove woodlands in the tidal zone).

These examples show that the fate of forests is often determined by phenomena in distant seas and oceans that the forest itself cannot do anything to counteract these effects.

Band two

The diversity of the world's forests is also clearly illustrated by another band extending from the north across Asia, Malaysia and Australia. In northern Siberia there are vast forest-tundras of larch and pine. Larches *(Larix sibirica,*

60

61

60 Siberian taiga.

61 Siberian taiga near the source of the Yenisei River.

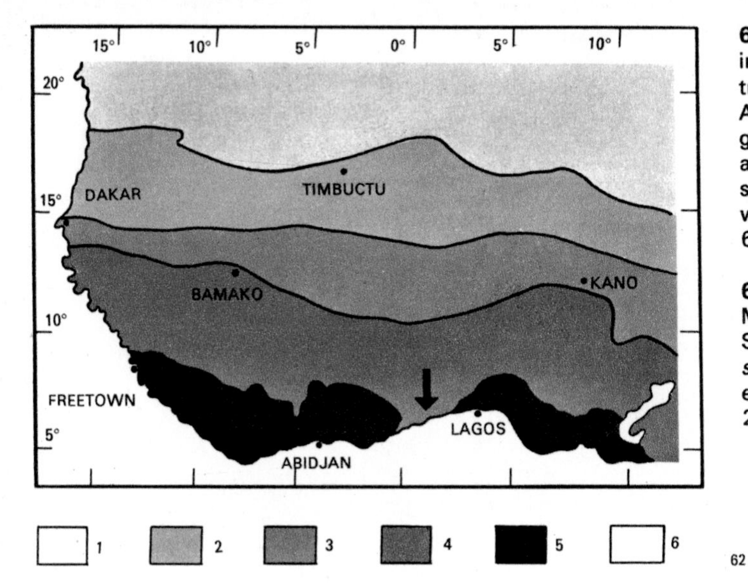

62

62 Dahomey Gap interrupting the belt of tropical rainforest in West Africa. 1. desert, 2. short grass savanna, 3. grass and shrub savanna, 4. tree savanna and savanna woodland, 5. rainforest, 6. mountains.

63 In the Tien Shan Mountains of central Asia Schrenk's Spruce *(Picea schrenkiana)* grows at elevations of 2,000 to 2,200 metres.

63

64

L. sukaczewii and *L. dahurica)* are unusual conifers in that they shed their needles during the freezing weather of winter and survive this cruel period with bare branches. During the brief summer they manage to grow new needles and produce sufficient material for growth and the building of new tissues. They incorporate the advantages of both deciduous broadleaved trees and sclerophyllous conifers. This makes it possible for them to grow in Siberia, and in regions with the most extreme climates such as Verkhoiansk and Oimekon, where winter temperatures often drop below minus 50 °C. Growth of trees in such an environment might be called a miracle but it is only one outstanding example of the vitality of the forest (Fig. 60 and 61).

The centre of Asia consists of vast steppes and semideserts, but mountain slopes are more often covered with coniferous taiga (Fig. 63) containing spruce, pine, larch and birch. In the Caucasus (Fig. 64 and 65) the dominant tree of the mountain forests is the Oriental Beech *(Fagus orientalis)* while in the warmer foothills it is the Spanish Chestnut *(Castanea sativa).* The world's highest mountain chain, the Himalayas, is bordered on the northern side by the high, practically treeless plateau of Tibet. On the southern side there are two na-

65

64 The Oriental Beech *(Fagus orientalis)* forms large forests in the Caucasus.

65 The Spanish Chestnut *(Castanea sativa)* once formed natural forests in the Mediterranean region, Asia Minor and the Caucasus; nowadays it is widely cultivated outside the area of its natural distribution.

66

66 A thriving fir forest with tree and shrub rhododendrons 3,600 metres above the Barun Khola Valley in the Himalayas.

tural forest limits: the upper or alpine limit and the lower or continental limit where the forest ends in the dry steppes of the foothills. The forests on the southern slopes of the Himalayas are thus clearly divided into several altitudinal belts. Highest up (above 4,000 metres) are the subalpine forests with the birch *Betula utilis* and large rhododendrons *(Rhododendron)* — (Fig. 66 and 67). Below these is the belt of coniferous forests with Indian Silver Fir *(Abies spectabilis)*, the Himalayan Hemlock *(Tsuga dumosa)*, several species of spruce *(Picea spinulosa, P. complanata* and *P. smithiana)* and the larch *Larix griffithii*, growing to the uppermost limit. Below 2,000 metres spreads the tropical evergreen forest, which besides oaks and members of the related genus *Castanopsis* includes trees of the tropical genera *Shorea, Terminalia, Ficus, Lagerstroemia, Bridelia,* and *Macarantha* as well as various palm and pandanus trees. The western end of the Himalayas has a drier climate and lacks the mountain mist forest. Stands of *Shorea robusta* occur at great heights and below them are extensive spreads of Deodar Cedar *(Cedrus deodara)*, Longleaved Indian Pine *(Pinus longifolia)* and various oaks *(Quercus incana, Q. dilatata)*.

Though there are now few forests in the overpopulated Indian subcontinent, dense tropical forests may still be found in the less accessible mountains of Kerala (Fig. 68 and 69). North-eastern China has deciduous broadleaved forests made up chiefly of deciduous trees of the beech family (Fagaceae). The forests in the southern parts of China, Laos and Vietnam are evergreen, composed of non-deciduous oaks and members of the genus *Castanopsis*. Also represented here in large numbers are members of the laurel and magnolia families.

67

68

67 Outside the avalanche path snow does not prevent the growth ot the vigorous Indian Silver Fir *(Abies spectabilis)* at an altitude of 3,500 metres in the Himalayas.

68 Even in the overpopulated Indian subcontinent there is still room for tropical mountain forests. Kerala, southern India.

69

From the seashores high up into the mountains, the Japan islands are covered with mixed broadleaved and coniferous forests (Fig. 70).

The Malay Peninsula (Fig. 71) is covered with dense forests broken only by inaccessible cliffs. The islands of Borneo, Java, Sumatra, Celebes and New Guinea were once covered, except for the high mountains, by continuous tropical rainforest which was felled only on the coast and round large human settlements in the interior. The trees of the upper layer of Malaysia's virgin forests are mostly members of the Dipterocarpaceae, Caesalpiniceae, Papilionaceae and Mimosaceae families.

Of all the continents Australia is the poorest as regards forest cover because the whole of the interior and the western part of the country lack regular rainfall. Those areas which have at least a brief rainy season are host to open sclerophyllous forests, with eucalyptus prevailing. In the mountains along the eastern coast there is a broad band of tropical forest which extends even beyond the Tropic of Capricorn. Semideciduous forests are also found in the north and south-west. In all these types of forest members of the genus *Eucalyptus* are the dominant element. The 600 species of eu-

69 Conspicuous elements in the structure of the virgin forests of southern India are the giant trees of the species *Actinodaphne madraspathana.*

70 Mixed forest on the coastal cliffs of Japan.

70

71 The vitality of tropical forest is evident in the way it covers steep cliffs without fertile soil in Malaya.

72 The Kauri *(Agathis australis)* is the dominant tree of New Zealand forests.

71

72

calyptus include trees adapted to all types of climatic and soil conditions, (mountain, thermophilous, drought-resistant and moisture-loving; species which will grow in poor soil, and so on). The savanna woodlands here also include acacias of diverse shapes, and numerous species of *Casuarina.* Neighbouring New Zealand is the home of vast deciduous as well as evergreen forests where kauri *(Agathis)* and southern beeches *(Nothofagus)* prevail (Fig. 72 and 73).

Band three

The forests of the Americas extend from the polar forest limit in northern Canada across the large mountain chains in the middle of North America, Florida and the Caribbean region to South America.

The polar limit is located at 60° N and coincides roughly with the July isotherm of 10 °C. The principal trees of the northern forests (Fig. 74 and 75) are the White Spruce *(Picea glauca)* and the Black Spruce *(Picea mariana).* Natural gaps, man-made clearings and areas destroyed by fire are reseeded by pioneer broadleaved trees consisting mostly of the Paper Birch *(Betula papyrifera),* Quaking Aspen *(Populus tremuloides),* Balsam Poplar *(Populus balsamifera),* and Red Alder *(Alnus ru-*

73

74

bra) (Fig. 76). Coniferous forests across the whole breadth of the continent from the Atlantic to the Pacific are the home of the Balsam Fir *(Abies balsamea),* Northern White Cedar *(Thuja occidentalis),* Tamarack *(Larix laricina)* and Jack Pine *(Pinus banksiana).* On the western boundary the coniferous forests include further coniferous species with a smaller area of distribution, confined only to individual mountain systems or restricted regions. These include the Pacific Yew *(Taxus brevifolia),* pines *(Pinus contorta, P. monticola, P. albicaulis),* larches *(Larix occidentalis, L. lyallii),* hemlocks *(Tsuga heterophylla, T. mertensiana),* firs *(Abies lasiocarpa, A. amabilis),* and the Sitka Spruce *(Picea sitchensia).* The coniferous forests in the east likewise have their particular species: the Eastern Hemlock *(Tsuga canadensis),* Red Spruce *(Picea rubens)* and large pines *(Pinus strobus* and *P. resinosa).*

On the west coast of North America the coniferous forests extend southwards along the slopes of the Rocky Mountains. The dominant

73 In the higher mountains of New Zealand there are still extensive stands of Menzies's Red Beech *(Nothofagus menziesii).*

74 In the northernmost part of the American continent the forest is made up of White Spruce *(Picea glauca)* and Black Spruce *(Picea mariana);* Alaska.

75 The only broadleaved trees to be found in the coniferous forests of northern America are the Paper Birch *(Betula papyrifera),* Quaking Aspen *(Populus tremuloides)* and Balsam Poplar *(Populus balsamifera).*

75

76

76 Alders also appear in the logged and burned sections of coniferous forests. The Red Alder *(Alnus rubra)*.

trees here are the Engelmann Spruce *(Picea engelmannii)* (Fig. 78), numerous species of pine *(Pinus monticola, P. lambertiana, P. flexilis, P. ponderosa, P. jeffreyi)*, firs *(Abies grandis*, Fig. 77 and 79, *A. concolor, A. lasiocarpa)*, Douglas Fir *(Pseudotsuga menziesii)* (Fig. 80) and, limited to the Sierra Nevada range, the famed Giant Sequoia *(Sequoiadendron giganteum)*. Coniferous and mixed forests also extend southwards along the ridges of the Appalachian Mountains, and contain, among other trees, balsam fir, red spruce and paper birch. Forests rich in conifers are also found at low altitudes round the Great Lakes (called lake forest by American foresters).

In the eastern part of the United States, south of the coniferous forests, is a zone of deciduous broadleaved forests. These are related to the deciduous forests of Europe and eastern Asia and contain many of the same genera. The deciduous forests of this area can be divided into three types. In the north-eastern part the dominant trees are the Sugar Maple *(Acer saccharum)* and American Beech *(Fagus grandifolia)*. Both grow well in shade, have a beneficial effect on the fertility of the soil and crowd out light-loving broadleaved trees. The most common subsidiary trees in this forest are various oaks *(Quercus borealis, Q. alba, Q. bicolor)*, the Yellow Birch *(Betula*

77

77 Crown of a Giant Fir *(Abies grandis)* native to the eastern part of North America.

78 Coniferous mountain forest with Engelmann Spruce *(Picea engelmannii)* dominant in the La Plata Mountains, Colorado.

78

alleghaniensis), American Basswood *(Tilia americana),* Tulip Tree *(Liriodendron tulipifera),* Sweetgum *(Liquidambar styraciflua),* ashes *(Fraxinus americana, F. pensylvannica),* American Elm *(Ulmus americana)* and Red Maple *(Acer rubrum).* A much-feared component of these forests is the Poison Ivy *(Rhus toxicodendron)* (Fig. 82). The conifers found here are either populations surviving from earlier, colder periods (particularly eastern hemlock) or pioneers reseeding open spaces which have dry and acid soil (mostly pines—*Pinus banksiana, P. strobus,* Fig. 81, and *P. resinosa).*

The deciduous forests further south along the Atlantic coast are composed of stands dominated until recently by the American Chestnut *(Castanea dentata)* together with various oaks, chiefly *Quercus montana, Q. coccinea, Q. borealis, Q. alba,* and in very dry places *Q. stellata* and *Q. marilandica,* the White Hickory *(Carya alba),* sweetgum, and others. Some sixty years ago practically all the chestnuts, which were one of the most important sources of wood, were destroyed by the fungus *Endothia parasitica* which was accidentally introduced to the area from Europe.

The western belt of the deciduous forest zone in the east is composed in great part of specialized oaks *(Quercus velutina, Q. coccinea, Q. phellos, Q. imbricaria, Q. macrocarpa)* and many hickories *(Carya ovata, C. tomentosa, C. ovalis, C. glabra, C. cordiformis).* These communities also include less specialized species common throughout the whole deciduous forest zone, mostly the sugar maple and American basswood, as well as many pines, like *Pinus echinata, P. taeda, P. palustris,* and *P. ponderosa* (Fig. 84), which cover burned areas and poor soils. Further west in Colorado, beyond the vast prairie region, the forests are composed of seasoned hardwoods, of the genus *Quercus, Prosopis* and *Cercocarpus,* for example (Fig. 83).

Florida at the south-easternmost tip of the United States is the domain of subtropical and tropical forests. Smaller sections of the state are covered with evergreen forest which includes typical tropical families. Found here, for example, are such species as *Bursera simaruba, Swietenia mahagoni, Ficus laevigata,* lianas and even Strangler Fig *(Ficus aurea).* A large part of Florida is covered with forests of Longleaf Pine *(Pinus palustris),* Slash Pine *(Pinus elliotii),* with the striking Saw-palmetto *(Serenoa repens)* and the small cycad *Zamia*

79

80

81

79 In fertile soil the trunk of a 70-year-old Giant Fir *(Abies grandis)* measures more than eighty centimetres in diameter.

80 The Douglas Fir *(Pseudotsuga menziesii)* is a conifer with characteristically cracked bark.

81 In New England the Weymouth Pine *(Pinus strobus)* invaded fields abandoned by farmers for more fertile lands.

82

82 Deciduous broadleaved forest outside Washington D. C.; Poison Ivy *(Rhus toxicodendron)*, the feared poisonous plant of American forests, climbs up treetrunks.

florida growing among the ground flora (Fig. 85). Along the coast of Florida are well-developed mangrove woodlands which show vigorous growth even beyond the Tropic of Cancer.

Much richer in species are the broadleaved and pine forests of the Caribbean islands, chiefly Cuba and Haiti. The variegated topography and lengthy isolation of the area has given rise to a very diversified forest flora, which includes both moisture-loving forms (particularly in the mountains) and forms that tolerate long periods of drought (tree savannas with pines and palms). A great many of the plants and animals in the forests of the Caribbean islands are endemics — species that grow nowhere else in the world.

Across the Caribbean are the vast savannas (llanos) of Venezuela and Colombia, consisting chiefly of small broadleaved trees of the species *Curatella americana* and *Byrsonima coccolobifolia.* Closed forest in this part of the world begins on the slopes of the Andes and southwards in the Amazon basin (Fig. 86). This vast area of rain forest is the largest continuous forest complex in the world.

The Amazon forest can be divided into several types. On ground which is above the level of regular flooding by the rivers, the forests contain huge trees of the legume family (Leguminosae) and members of the mahogany, nutmeg (Myristicaceae), Rubiaceae, Compositae, laurel and Flacourtiaceae families. Also growing below the canopy are many palms and smaller trees, one of which is the Cacao *(Theobroma cacao)* which yields a product popular the world over.

In the vicinity of rivers in the Amazon basin there are two types of flood plain forests. Alongside rivers that carry large amounts of fertile loam down from the mountains (so-called white rivers) the flood plain regularly receives rich alluvial deposits on which grows a luxuriant forest called varzea. The trees here must be able to survive immersion in water for several weeks, and at the same time make the best possible use of the rich deposits for rapid growth. In these forests are found the largest trees in South America. The Silk Cotton Tree *(Ceiba pentandra),* for instance, grows to a height of fifty metres and has large flat buttresses at the base which look like rock walls (Fig. 87). Another type of floodplain forest is the igapo, found alongside so-called black rivers like the Rio Negro. The water that flows in these rivers comes from flat peat basins. It is

83

84

83 Remnants of thermophilous forest at the foot of the Rocky Mountains in Colorado.

84 Dwarf stands of Ponderosa Pine *(Pinus ponderosa)* in old lava fields; Sunset Crater, Arizona.

85 In the south-eastern parts of the United States stands of Slash Pine *(Pinus elliotii)* with understorey of Saw-palmetto *(Serennoa repens)* grow in places frequently damaged by fire.

86 On the banks of a river the tropical rainforest in Ecuador is veiled by a mass of smaller woody plants and herbs, including tree ferns.

85

poor in nutrients, and floods the land for long periods, so the trees that grow in the igapo forest are only those which tolerate lack of oxygen in the soil. Besides moisture-loving dicotyledonous trees, certain palms are also exceptionally well adapted to this environment.

The Andes, of course, are covered with a different type of forest, particularly at higher altitudes where the high mountain climate with large fluctuations between day and night temperatures and the resultant mists have a marked influence on the vegetation. In such forests there are increasing numbers of *Podocarpus* species *(P. respigliosi, P. montanus, P. oleifolius).* In Colombia there are also oaks and hickories at high altitudes. Highest up in the Andes are groves of the small tree *Polylepis sericea* of the rose family (Rosaceae), which is found even at elevations above 4,000 metres and so equals the record for height of the Himalayan birch *Betula utilis.*

South of the Amazon forest complex, in southern Bolivia, Paraguay and northern Argentina, spreads the extensive lowland plain called Chaco, which, except for the land round the rivers, is covered by dry, deciduous forest. The composition of the forest varies according to the soil and ground water conditions but it is mostly quebracho forest with *Schinopsis* and *Aspidosperma* species prevailing. The common characteristics of these trees is their hard wood which 'breaks hatchets' (quiebra

86

87

hacha in Spanish—hence quebracho) and which is used in construction work, for railway sleepers, and for its excellent tannin. Also found in the Chaco are woods composed of leguminous mesquites *(Prosopis)* which grow in salty conditions.

Still further south lies the great Patagonian plain and its pampas. Forest is present on the slopes of the Andes and in sheltered places in the southernmost tip of the South American continent and neighbouring Tierra del Fuego. These forests are partly deciduous and partly evergreen, the prevailing trees being species of *Nothofagus,* monkey puzzles *(Araucaria)* and *Libocedrus.* The forests of Tierra del Fuego are composed of *Nothofagus dombeyi* and *N. betuloides.* These are supplanted in the mountains at the upper forest limit (which is a mere 200 to 300 metres above sea level here) by *Nothofagus pumilio.*

In this general survey of the forests of the world no mention has been made of the special types occurring on unusual soils and in extreme climatic conditions. Forests that grow in bogs, peat moors, the tidal zone, stone rubble and sands will be discussed in Chapter 3. One thing, however, is certain—the earth's green cloak of forests was sewn by a very imaginative tailor.

87 Buttresses of a Silk Cotton Tree *(Ceiba pentandra)* rise like rocks from the undergrowth of the tropical forest.

88

88 Stacked wood.

FOREST AND MAN

Were it not for the spread of the human race there would be far more forests on the earth, and in many places they would be very different in appearance. Man has greatly changed the surface of the earth both by destroying the forests and by changing them to meet his needs. Over periods of thousands of years geographical factors determined whether a territory could or could not be covered with forest but now to this has been added the human factor, and man's influence on the forest can be observed in every part of the world.

In densely-populated Europe, forests are even on the retreat in the high mountains (Fig. 89) where agricultural enclaves are spreading round solitary houses, small hamlets and larger communities. On the one hand it is necessary to have more fields and pastures, but on the other, there is a need for the products of the forest—timber, forest litter, fertile soil and forest animals. Even in those places where there has been a strict division between land allotted to farming and that left to forest, the appearance of the forest is changing. Foresters themselves fell trees on large tracts (Fig. 90 and 91), creating rigid, artificial forest margins in the landscape, changing the composition of the forest by restricting the number of species growing there and developing highly productive organized plantations. People have become accustomed to a landscape having clear swathes cutting straight through areas of dense forest and with deforested areas around human habitations. Most people like an open landscape—but with a forest not too far off. In some parts of the world a partially deforested area may resemble the English parkland landscape, though on a larger scale (Fig. 92). Generally, however, normal deforestation of the countryside is a far cry from the planned and creative approach of the founders of natural parks.

In regions where the development of human civilization was intensive over thousands of years the results of man's exploitation of the forests are irrevocable (Fig. 93). In the Mediterranean region (Fig. 94, 95 and 97) the original broadleaf forests and pine woods have almost completely disappeared and all that remains is scrubby vegetation—maquis and guarigue—of no economic value and only of negligible importance in holding the soil. Erosion has washed away most of the fertile soil and the return of the high forest is either a matter of time measured in centuries or else of costly recultivation involving the bringing in of fertile soil. Despite this lesson, the exploitation of forest continues unabated in other parts of the world. In the case of deeply weathered tropical soils on steep slopes (Fig. 98), destruction can be complete within a matter of years.

The forest as shelter

The history of man's relationship with the forest goes far back into the past and is very complex. Man's predecessors were adapted to life among the treetops. To this day our limbs, senses (in particular our sight), and our digestive system bear witness to our physical ties with the forest and to a vegetarian diet. The locomotor system and senses of our ape ancestors, the ramapithecene apes, had already become adapted to life in the open savannas and on the shores of lakes and rivers. Walking upright on two legs instead of on all fours as well as the modification of the hand were physical adaptations of the earlier savanna ape. Most important was doubtless the development of mental capacity. The move from the safe and practically constant environment of the tropical forest to the changing environment of the savanna was probably the basic impulse that led to the perfection of the brain and the development of social organization which culminated in *Homo sapiens.*

Though man lived in the grassy savannas, he still returned to the forest for his food, fuel, building material and medicines. Early man liked to live in clearings and open savannas but he needed to have the forest at his back. Paths always led from human habitations to the forest and frequently, either of his own free will or because he was forced to do so by the pressure of enemies, man went back into the forest depths.

Today's paths, roads, railways and navigable rivers also lead to forests. Newly built roads provide access to areas of virgin forest where man has never before set foot. There, either man's enterprising and commercial spirit

89

89 An inviting landscape in the Italian Dolomites made up of the remains of a plantation, pastureland, fields and dwellings.

90 A new clearing shortly after the removal of felled beeches.

91 A large clearing in a mountain spruce forest.

90

gains the upper hand and the original dirt path is soon turned into an asphalt road that facilitates the removal of the forest's products; or else the forest is successful in fending off the onslaught, its vegetation covering man's paths and obliterating all traces of human activity. More often the presence of a path through a forest becomes the starting point for the ousting of trees from the landscape. Along trails in the wooded valleys of the Himalayas (Fig. 100 and 101), as alongside paths the world over, there are corridors of felled trees—the result of centuries of man's habitation. Wood from trees adjacent to such paths is often transported great distances because fires and the construction of human dwellings require a constant supply of timber. The first dwellings in forest do not require much construction material (Fig. 102), particularly if, in this modern day and age, use is made of tents or even trailers (Fig. 103), but soon permanent settlements—first hamlets and then villages and towns—spring up around the paths and these give rise to the rapid spread of treeless tracts for use as fields and pastures (Fig. 105).

The relationship between forest and man changes with the development of human society. Forest peoples on the level of fruit gatherers and hunters of wild game have made little impact on the forest proper. Ethnographers have described in detail the life of Pygmies in the virgin forests of the Zaire basin, South American Indians in the Amazon forests and

91

the Punan tribes of Borneo's forests. Such tribes take from the forest only what they need for subsistence and at the slightest sign of a more radical change in their environment they move to a different territory. Their modest property is easily carried to their new home and their simple dwellings are easily built anew with the materials provided by the forest.

Most forest-dwellers in various parts of the world do not live in the heart of the forest under the closed canopy of the trees. Many tribes have made their way to the forested regions bordering rivers and have settled in clearings by the water where they are able to enjoy not only the bounty of the forest but also the facilities the river provides for washing, fishing, transport and travel. A village situated on a riverbank may be at some disadvantage, however, for it is more susceptible to attack by enemies, it may be infested by troublesome insects and it may be more subject to epidemic diseases. That is why in all parts of the tropics communities such as the Indian tribes of South America can be found deep inside the forest in small natural or man-made clearings (Fig. 104).

Forest Indians build large communal dwellings for whole clans with sometimes separate lodgings for the men. Such dwellings are often raised structures resting on stilts, to protect the inhabitants from floodwater and troublesome animals. Only a few crops, such as sweet potatoes (batatas), manioc, tobacco and various plants of the cucumber family (Cucurbitaceae) are grown but many foods, delicacies, medicines, narcotics, poisons, dyes and

92 The remains of a flood plain forest in the Dyje River valley. Czechoslovakia.

A farmhouse in a clearing in the heart of wooded country in eastern Australia.

A felled, grubbed and burned clearing made ready for growing corn in the tropical forest of West Africa.

An African farmer fights the forest with fire; the hardy West African Oil Palm *(Elaeis guineensis)* continues to grow on the burned ground.

93 Area covered by forest in various countries (as a percentage of the total area of the country).

94 Shrub forest and Italian Cypresses *(Cupressus sempervirens)* on the Dalmatian coast, Yugoslavia.

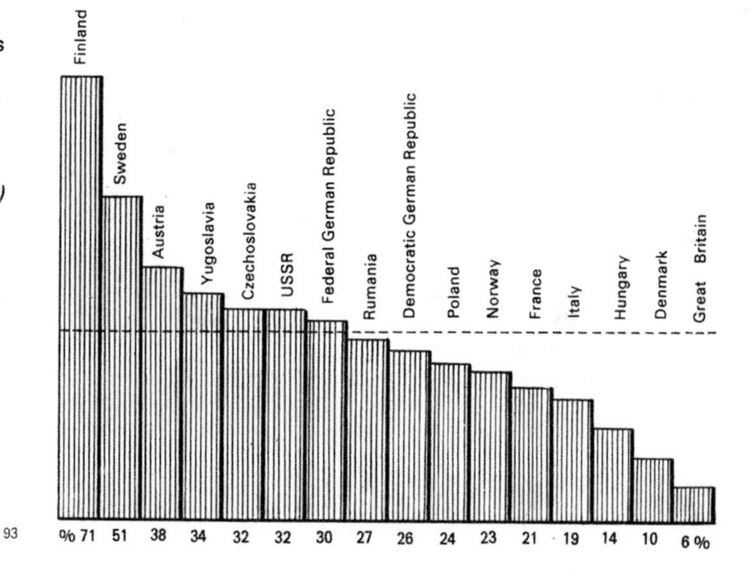

93

94

material for household needs are gathered from the forest. Many tribes live by hunting, their chief weapon being a blowgun and poisoned arrows. When the soil is exhausted by the cultivation of crops, or when the tribal chief dies or some other disaster befalls them, forest Indians move to another spot and start all over again, beginning by burning and felling trees to make a clearing.

There are, however, a few rare instances of isolated groups who make their homes in the heart of the closed forest. The Semang and Sakai tribes, for example, live in the impenetrable forests of Malaysia. The most typical examples of true forest dwellers, however, are the Pygmies of Africa, an ancient branch of the Negro race. At some time the Pygmies took shelter in (or were driven into) the forest and in the isolated and equable environment of the tropical rainforest they remained a Stone Age people. They lack a common language and have a simple culture. It is quite possible that the Pygmies are closely related to the Bushmen and Hottentots of the neighbouring savannas and steppes of the Kalahari region and that their small stature is the result of their lengthy evolution in the forest and the process of natural selection. Pygmies are constantly on the move within the forest, never staying long in one place, so they build only simple dwellings of branches and palm leaves. They are first and foremost hunters, using small bows and poisoned arrows. Their catch usually consists of small forest antelopes, monkeys and birds, though occasionally they may get a large antelope. They make good use of all available edible fruits and roots, and on rare occasions also obtain foods from the farms of their current neighbours. The tropical forest may have been the cradle of man's ancient predecessors but it is definitely not an environment conducive to the evolution of a 'developed' society.

Savanna woodlands and grass savannas with scattered trees were, however, far more conducive to such a development. The fluctuating climate of these regions, and particularly the marked changes in the aspect of the landscape during the course of the year caused by the periodic growth of grasses and trees in re-

95 A barren limestone region of Yugoslavia with shrubby oaks *Quercus pubescens* and hornbeams *Carpinus orientalis.*

96 A deforested part of Macedonia, Yugoslavia.

95

96

97

lation to the dry and rainy seasons, forced the human race to a higher level of organization. This led to the production of more ingenious tools and to the perfection of various means of communication, especially speech. Even such primitive tribes as the Bushmen and Hottentots have an excellent practical knowledge of plant and animal ecology and phenology. In time, most primitive peoples in the savannas became farmers and herdsmen. A great ally in their struggle against the forest when they cleared ground for fields and pastures was fire. In the dry season vast tracts of land could be cleared by burning and so the borders of the closed forest slowly receded from the savannas. Simultaneously man put to good use the isolated islands of grassland which occur on poor soil in the midst of vast forest complexes. In the Zaire River basin such islands are called esobe.

Beyond the steppes and savannas, on the southern edge of the Sahara and in the Kalahari region, the forest recedes along a broad front and there is a rapid reduction in its spread. In East Africa the struggle with the forest has been going on for several million years as testified to by finds of the first species of Man—*Homo habilis* and *Homo robustus.*

The decline of the forests

Man's influence on the forest became more pronounced during the Neolithic Period, when he began to rear stock and grow crops in fields. Soon he learned how to store and transport water and also how to preserve food. Thereafter the spread of fields and pastures forced the forest to retreat with a rapidity proportional to the growing population and the higher level of social organization. In the temperate zone the primitive farmer renewed the fertility of his fields by letting the ground lie fallow. In the tropical regions the farmer abandoned his fields, if not after the first harvest then after the second, and laboriously cleared new land. Fire was the main instrument used by Neolithic man to aid him in his task. The pasturing of cattle also forced the forest to retreat and the consumption of wood for fuel and building hastened the process still more. In regions with conditions congenial to agriculture forested areas were soon reduced to as little as thirty per cent of the original area.

How rapidly forest country can be transformed into a sparsely wooded region is well demonstrated by the colonization of New Eng-

97 The total deforestation of Albanian hills was caused by centuries of grazing and logging.

98 A hill devastated by burning and grubbing outside Zipaquirá, Colombia.

99 A newly-cut path through the Ankasa Reservation, West Africa.

98

99

100

100 The last of the forest in the Arun River Valley in the Himalayas will be gradually carried away by bearers of fuelwood.

101 The forest of Morinda Spruce *(Picea smithiana)*, Himalayan Fir *(Abies spectabilis)*, Deodar *(Cedrus deodara)* and Bhutan Pine *(Pinus wallichiana)* is clear-cut along the Himalayan mountain trail.

102 A simple but efficient hut in the tropical rain forest, West Africa.

101

102

land in the north-eastern United States. Near the beginning of the eighteenth century some parts of New England were inhabited by settlers from Europe. Within a mere 150 years the age-old forests were reduced to a third of their original size. Examples have been cited, however, of western areas of the United States which were quickly won back by the forest when man abandoned his fields and pastures for better and more fertile lands, but this is the exception rather than the rule.

The size and disposition of the forested areas of the modern world has been in great measure predetermined by the settlements of former times when the forest remained only in places which were unsuitable for agriculture,

103

such as steep, rocky slopes, poor sandy areas and in the flood plains of large rivers (Fig. 109 and 110).

To make the forest give way to his needs man equipped himself with suitable tools. The axe was his first major weapon against the forest giants (Fig. 106) for with it he could fell both softwood conifers and hard-wooded tropical giants. It was much later that he began to use the saw. The two-handed lumberman's saw (Fig. 107) was used until after the Second World War when it was replaced by new and efficient power saws. Thick tropical trees with their spreading buttresses always presented a problem for the tropical farmer or lumberman. Where possible he built a simple scaffold and chopped or cut the tree above the buttresses (Fig. 108). Particularly large giants were killed either by the removal of a ring of bark around the trunk (girdling) or with fire at ground level, and then they were left standing in the field. Modern technology has produced an improved method of killing trees with the introduction and use of specific poisons called arboricides (Fig. 112).

Time has never been charitable to the forest. Terrain which had hitherto been safe from man's encroachment became progressively more vulnerable as he acquired the skills necessary to construct dwellings on and cultivate, for example, steep slopes and swamps. In many mountain ranges flocks of sheep and goats completed the work of devastation. Their ability to climb to even the least accessible places and the fact that they have no particular likes or dislikes when it comes to nibbling at trees, prevents both natural as well as planned regeneration of the forest. In many

103 A caravan is taken to the end of the trail and then a snowmobile is used to penetrate the untracked wilderness of an Alaskan forest.

Mediterranean and Balkan localities, the land became totally unfit even for agriculture.

The products of the forest
Food

Forest plants and animals have always served as food for man in all parts of the world. The sweet fruits of trees and herbaceous plants have always been gathered by the forest dwellers. Juices like the sap from the vascular bundles of palm trees have been fermented to make wine (Fig. 113). Starchy tubers and the starchy pith of palms have provided more substantial food and juicy berries have always been a popular item of man's diet. Though cacao was found to be a delicacy at a relatively late date (Fig. 118), the fruits of forest coffee trees were harvested in ancient times (Fig. 119). Mushrooms are gathered for food in

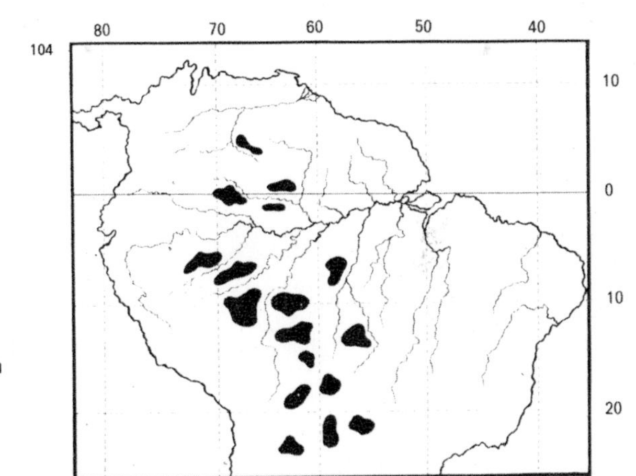

104 The distribution of native Indian tribes of hunters and food gatherers in the forests of South America.

105 Typical layout of a village alongside a trail in the African bush.

105

106

107

106 Two loggers fell a spruce with axes.

107 Felling a tree with a handsaw.

108 Framework erected prior to cutting a tropical tree above the buttresses; Ghana.

most forest regions of the world though they are most prized in the temperate zones (Fig. 121). They have been called the meat of the forest from time immemorial though until comparatively recently they were thought by some to be a delicacy with no other value. It has been proved, however, that mushrooms contain, among other things, valuable amino acids, and it is not unusual for some forest species to be home-grown (Fig. 120.)

Even though man's distant ancestors were mostly vegetarians, all known forms of man ate meat. Excavations in East Africa have yielded remnants of fishes, molluscs and amphibians which were consumed by *Homo habilis.* Round the settlements of present day forest dwellers there are always rubbish heaps that reveal how zealously man hunts and gathers the edible forest animals. The protein thus obtained is an essential component of his diet. One of the simplest means of satisfying this need for protein is to gather forest molluscs which are not only tasty but are also a substantial source of food. The large tropical snail of the genus *Achatina* yields five to ten times more meat than the garden snail *(Helix pomatia).* Baskets filled with these snails are a familiar sight in the village markets of the remote forest. The natives also gather certain crabs that live in forest swamps, as well as grasshoppers and locusts. Frogs, turtles, lizards and snakes are standard items in their diet.

Birds and mammals are the commonest source of protein in forest throughout the world. It is only a matter of local custom whether pigeons and turtledoves, parrots, songbirds, or game birds such as the capercaillie or the black grouse are hunted in a given region. In the forests of today's developed nations special care is devoted to the breeding of various species of game birds and the shooting season is regulated by strict rules. Forest mammals, however, are the most popular source of game. Mammals of all sizes, from small rodents to large beasts of prey, ungulates, elephants and monkeys, are hunted in the depths of the tropical forests. Hunters of the Amazon forests take as much delight in catching a tapir as do hunters of Africa's forests in capturing a Colobus Monkey or a European hunter in shooting a red deer (Fig. 114, 115 and 116). For forest dwellers it is the meat that is most important but the European hunter prizes the trophy. In some instances even primitive peoples set greater store by the skin or teeth of the captured beasts. Trophies of animals of the cat family are particularly prized and are important ornaments and sym-

108

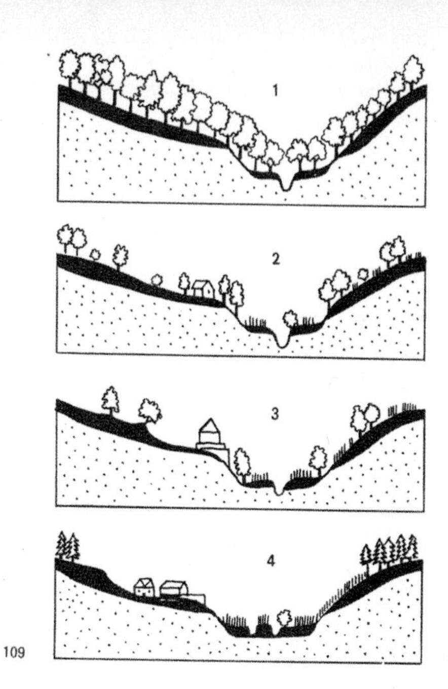

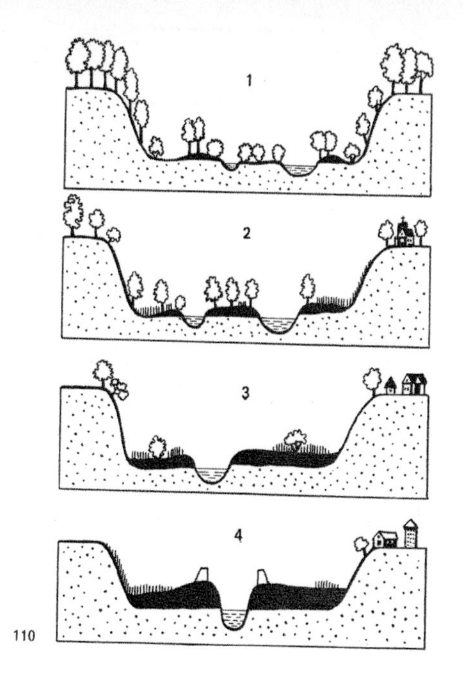

109 The course of deforestation in a densely populated landscape on the upper reaches of a European river. 1 — before Christ, 2 — 1000 AD, 3 — 1800 AD, 4 — 1900 AD

110 Gradual changes in a landscape on the lower reaches of a central European river. 1 — before Christ, 2 — 1000 AD, 3 — 1800 AD, 4 — 1900 AD

111 Efficient machines now do the heavy work in logging operations.

bols in all parts of the world. Leopard trophies are an essential part of the trappings of secret leopard societies in equatorial Africa whose members include witchdoctors, magicians and sometimes even swindlers.

Medicines

The forest is also a boundless source of traditional as well as modern medicines for both man and domestic animals. The drugs are obtained from flowers, fruits, leaves, bark, wood, roots and tubers. Primitive peoples have acquired a wealth of knowledge and experience that has not yet been objectively evaluated by medical and veterinary science. The traditional drugs of European and North American forests are fairly well known, but in most cases tropical plants have not been investigated to find the substances they contain.

Herb healers and medicine men jealously guard their secrets which is why in many tropical countries traditional methods of healing are successfully practiced side by side with modern medicine.

112

113

112 Killing a tree with arboricide.

113 Harvest of palm wine tapped in the crown of a *Raphia hookeri* palm; Sierra Leone.

114

115

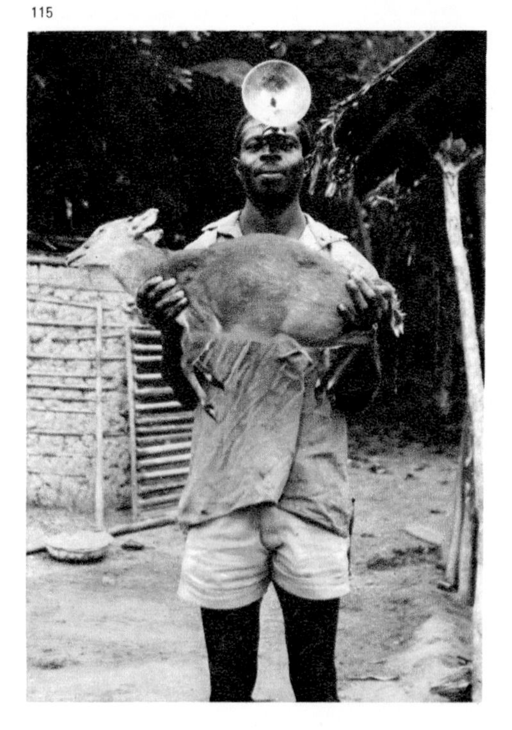

Some families of forest plants such as the Apocynaceae, mulberry and legume families contain numerous alkaloids. The roots of *Rauwolfia serpentina* of Indo-Malaysia contain forty different alkaloids, including reserpine, which has outstanding sedative properties and is also used in the treatment of hypertension (high blood pressure). The extracts of many plants have stimulating properties. Members of the genus *Cola* yield an extract used as the basic substance for many popular non-alcoholic beverages. Plant toxins, and natural antitoxins that can be used to counteract the effects of dangerous substances, continue to be of practical importance to the life of many peoples.

Building materials

The forest was, and still is, a rich store of materials used by man for building his dwellings. Small poles, pole timber and roofing materials are available in great abundance in the tropical forests. Bamboo and the light, pliant and firm leaf stalks (rachises) of tropical palms make ideal construction material. In West Afri-

114 A poacher with his haul of Colobus Monkeys *(Colobus* sp.*)* from the African forest.

115 A hunter has captured a forest duiker *(Cephalophus* sp.*)* with the aid of a carbide lamp.

116 A European forester with a Wild Boar *(Sus scrofa).*

117 The rapid deforestation of a landscape in Venezuela following the construction of a main road: left, in 1947, right, in 1952 following the construction of a second road.

116

117

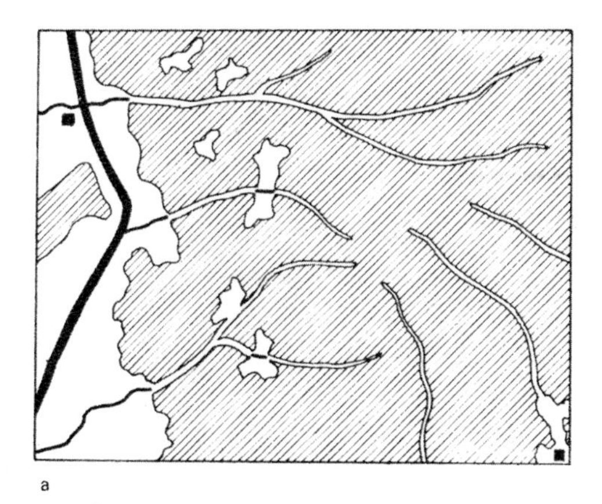

a

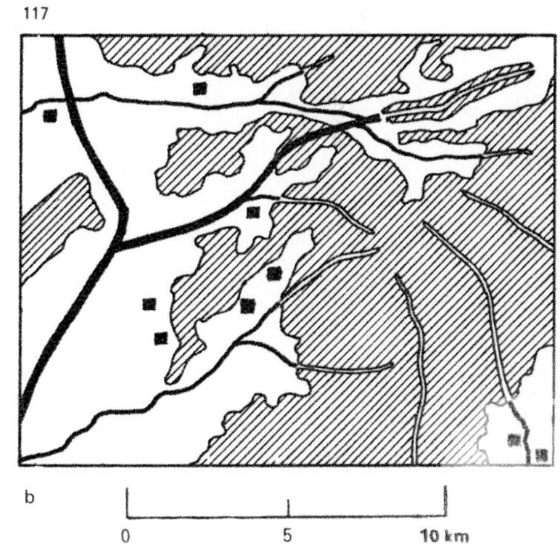

b

ca (Fig. 122 and 123) the natives construct the framework of their houses from the leaf stalks and rachises of the palm *(Raphia hookeri).* They plaster the framework with mud and roof it with braided palm leaves. Huts in the Amazon region are usually made entirely of wood, and are often raised above the ground on stilts (Fig. 124). Very similar dwellings on stilts standing in water in the well-known village of Ganvier in Dahomey are made of poles cut from mangrove trees (Fig. 125) as are dwellings on Africa's east coast. Construction material in tropical areas is chosen for its resistance to damp, termites and rodents.

118 Fruits of a Cacao *(Theobroma cacao)* grown in the shade of the virgin forest; Ghana.

119 Drying the berries of the coffee tree *(Coffea robusta)* in the mountains of Togo.

118

119

A field of rice and corn in the heart of a tropical rainforest in Peru.

Clear felling on the banks of a river valley prior to flooding by a reservoir.

120

In coniferous forests of the north the ideal construction material was always the trunks of conifers. These are quite durable and easily processed. With the aid of simple tools they can be made into a warm hut in which it is even possible to pass the winter (Fig. 126 and 127). Conifers were formerly used to build whole villages in the Carpathians (Fig. 129) and moss was stuffed into the cracks between the logs to help make the cabins weatherproof. In the mountains of central Europe, sturdy cottages were also made mostly of wood, as were many picturesque churches (Fig. 130), built without a single nail. These churches were long-lived and some can still be seen in Scandinavia which are more than 500 years old.

In the forests of both the tropical and temperate regions, of course, there are buildings made basically of stone but which have close associations with the surrounding forest. The magnificent stone structures of the ancient Mayan Empire in Mexico (Fig. 128) could have been built only in a society that exploited the wealth of the surrounding forests. Castles

120 In Japan the edible polypore *Lentinus edodes* is grown on stacks of wood.

121 Fruiting body of the edible mushroom *Gyroporus cyanescens* growing in a mixed forest.

121

122

122 An African village in the heart of the tropical rainforest.

123 The framework of the walls of this dwelling is made of the rachises of the *Raphia hookeri* palm tied together with lianas; West Africa.

124 Construction of a hut in the Amazon region of Ecuador.

123

in Finland's lake country (Fig. 131) had walls of stone but the roof trusses, gates and interior furnishings were made of wood. In Finland's modern housing developments, pine forests are used as a scenic background to the urban development (Fig. 132).

Transport

As well as providing material for building homes, the forests have also throughout the centuries supplied materials for progressively more sophisticated forms of transport. The large Arun River in the Himalayas is spanned by a suspended bridge built solely of light poles and lianas, (Fig. 133) and the lianas of the Araceae family are still commonly used in tropical regions as strong ropes for securing all manner of things. The old covered bridges of New England rest on wooden arches that bear an enormous load. An old cartwright in a small village on the shores of the Arctic Ocean in northern Norway was still able to make a wooden cart without a lathe and complex fittings not so long ago.

The hard wood of broadleaved trees was, on account of its durability, invaluable for the construction of ships, bridges, sluicegates, waterwheels, wagons and railway sleepers. That is why the oak forests in the Mediterranean region and the British Isles were so decimated. In Europe, the beech was widely used for making railway sleepers, particularly when the problem of effective impregnation was solved, and in South America the Quebracho forests were distributed throughout the continent in the form of railway sleepers. In North America, industry always had a greater choice of hardwoods, most popular being various oaks, hickories, American chestnut (virtually destroyed by an introduced fungus disease), yellow birch.

The use of trees by tropical peoples

The primitive peoples of the tropical forests have never needed much in the way of clothing, their garments being little more than of a ritual nature. Low-slung skirts, simple slippers and a wide-brimmed cover for the head to protect it from the scorching sun can be easily woven from the inner bark of many trees and the leaves of numerous palms while other garments can be fashioned from the skins and feathers of forest animals. The forest provides wood, bark, leaves, lianas, fruits and seeds for the making of various articles of daily use such as furniture, cooking utensils, fishing tackle and hunting weapons.

Fuel

Since time immemorial man has used large quantities of wood for fuel. The fire in various stoves, ovens and fireplaces had to be kept continually burning because it was necessary to prepare hot food and also keep warm in cold weather. Later, wood was used not only as domestic fuel, but also in steam boilers which supplied the needs of developing industry and expanding transport (Fig. 137). Not so long ago river and lake steamers plied from landing stages piled high with wood that fu-

124

125 Dwellings on stilts in Ganvier; Dahomey.

126 Hunter's shelter in the wild Siberian taiga.

127 Log cabin in the far north of Alaska.

125

126

127

129

elled the large boilers in their holds. Large quantities of wood were processed into charcoal by burning in charcoal piles (Fig. 138). Not until recent times, and then only in developed nations, was wood replaced by coal and oil. The world's reserves of fossilized fuel, however, are being rapidly depleted. While new sources of energy are being sought, there are those who voice the opinion that the only solution to the energy crisis of our day is the efficient use of a renewable source of energy—fuel from trees and shrubs. One energy expert calculated that a return to wood as fuel could supply the entire requirements of modern society. What, however, would then be the fate of the forests?

Timber in industry

Wholesale destruction of forests was also brought about by the development of industry in various parts of the world, and the concentration of people in large cities. Only those types of wood which could be processed economically by the currently available methods were used. In the northern hemisphere conifers were the trees traditionally exploited so many coniferous forests had already been decimated. Even today conifers are no less important. In Europe it is the wood of the Norway spruce and Scots pine that is most used. In North America there are many more coniferous species to choose from but various pines, spruces and firs again head the list.

The wood of forest trees is not only a good construction material, but also a valuable raw material for the paper and chemical industries. Besides the main components—cellulose and hemi-cellulose—wood also contains a number of important extractible substances, as for instance ethereal oils, volatile carbohydrates, acids and fats, natural dyes, tannins, soluble

128 Mayan temple covered by vegetation in a tropical forest in Mexico.

129 Cabins made of logs of Norway Spruce *(Picea abies)* and covered with shingles of the same wood.

130 Wooden church built without a single iron nail.

130

131

saccharides and glycosides (starch, glues, slimes and pectins), mineral salts and albumins. Many forest peoples, who have no access to other natural salt sources, are dependent on the burning of wood as their only source of salts. Tannins obtained from wood and bark have been used for dressing skins since time immemorial. Starch extracted from the pith of certain tropical palms (sago palm) is a valuable source of food for the nations of southeast Asia.

The largest amount of wood, however, is used in the production of cellulose, which today represents the basic raw material in the paper industry. Not all trees, however, are

131 Stone castle in Finland's lake country.

Logging in the mountain forests of central Europe.

Flocks of grazing sheep in the Carpathians.

Shrubby forest of chaparral-type with the cactus *Backebergia militaris* (central Mexico).

A giant bulldozer easily undermines vast stretches of Alaskan taiga.

Bearberry *(Arctostaphyllos* sp.*)* grows in abundance in the northern forests on the edge of the tundra.

equally suitable for this purpose. Best of all are young conifers (firs and spruces), but certain broadleaved trees, as for instance maples, limes, willows and poplars, are also used with good success. The trees of tropical forests are in greater part unsuitable for this aim, and therefore artificial plantations of coniferous and broadleaved trees good for the production of cellulose are being set up in these regions.

The destruction of the forests
Fire

Man often burned large tracts of forest either without purpose, for petty reasons, out of ignorance, or because of his imperfect technology. Felling a forest to clear land for agricultural purposes is often thought to be a sound economic idea, with the wood being immediately used for construction, fuel, or for the products it yields. However in many situations this is by no means the case. Often, man has also needlessly burned trees as well as forest litter and automatically destroyed thousands of woodland creatures which make their homes in the forest soil or forest canopy.

In savanna woodlands the dry season was the best time for burning, while in the forests of the temperate zone man chose early spring, when the previous year's dry grass had mouldered and become combustible. In the damp northern forests, as in the tropical rain forests, man knew how to make the necessary preparations. First he girdled a number of trees or cut down smaller trees among the undergrowth, and when the larger trunks had dried he set the entire forest on fire.

Trains pulled by steam locomotives at one

132 A modern housing development set amidst old stands of pine; Finland.

133

time caused great damage to forests throughout the world and even today, in some places, sparks from their stacks set fire to undergrowth, dry grass and nearby trees. In many countries attempts are made to prevent this by digging protective ditches on either side of the track but in all probability this danger will be a thing of the past only when all steam locomotives have been replaced by diesel and electric engines.

In those parts of the world where villages and towns are still built mostly of wood the forest's existence is also seriously threatened by fire that breaks out within the confines of such a village. A burned-out village must be rebuilt and the material for reconstruction is once again taken from the forest. Shortly after a fire the surrounding forests are invaded by teams of loggers with transport vehicles of all kinds to bring in timber for restoring the hamlet. Then follows the work of construction —the squaring of logs, cutting of planks and wooden shingles, and the actual building. When all is finished, disaster still lurks around the corner and another fire may be the cause not only of further human misfortune, but may

134

133 Suspended bridge across the Arun River in the Himalayas; Nepal.

134 Manufacture of wooden dolls on a hand-operated lathe; Japan.

135

be responsible for the disappearance of yet more forest from the face of the earth.

The demands of industry

In the eighteenth century Europe and North America began to show increased interest in the wood of tropical broadleaved trees. This was brought about by the enthusiastic reports of pioneering overseas expeditions made by naturalists, particularly those of Baron Alexander von Humboldt to the tropical forests of Latin America. Reports concerning the extraordinary beauty and excellent characteristics of tropical timbers were often accompanied by samples of useful objects or souvenirs of great value. Demands for mahogany, ebony, cedar, rosewood, satinwood, marblewood, balsam and teak flooded the world market. It was some time before the local names of woods were associated with the botanical species to which they belonged; however, it was later found that in different tropical regions the same commercial name was used for species that were quite different from one another.

The developing industrial nations, for instance, were enthusiastic about the wood called mahogany. True mahogany is from the giant South American tree *Swietenia macrophylla*. The wood is not only extremely beautiful and durable but is also easily worked and

135 Tombstone hewn from the wood of tropical trees; Malaya.

136 A sabre-shaped tree is hewn for an allegorical float in a religious festival; Nepal.

136

polished. In the days when there were no light metal alloys or plastics, mahogany was a priceless material for lathe turning, wood carving, furniture making and machine building, for it was used to make important parts for many machines and instruments. No other comparable material had such outstanding characteristics. Later, however, other trees that yielded so-called mahogany of equally good quality as the South American species were discovered in other parts of the world, e.g. the African mahogany *(Khaya ivorensis)* of the Meliaceae family and the Australian red mahogany which is a species of eucalyptus of the Myrtaceae family.

Tropical woods are noted for the many varied characteristics which cannot be replaced by any other material and which fulfil the varied demands of the many branches of industry as well as the diverse tastes of designers and consumers. Their range of colours includes even such unexpected hues as pure black, blood red, pink and bright yellow. Tropical woods exhibit countless variations in structure depending on the cut. Even more remarkable are their physical characteristics. Corky woods like that of the Ambach tree *(Aeschynomene elaphroxylon)* have a relative density of only 0.1, while at the other end of the scale, so-called iron woods have a relative density of 1.0.

In many countries the sale of timber provides a substantial proportion of the national income. Countries like Canada and the Soviet Union, whose territory embraces the greater part of the northern coniferous forest, have developed an extensive timber industry which has an assured market throughout the world. Logging is also very important in tropical areas, such as West Africa, Brazil and Indonesia

137

138

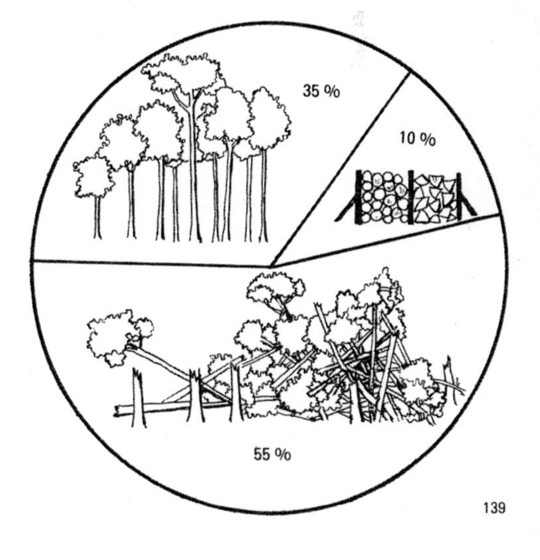

139

137 For decades steam boilers were fired mainly by wood; Sierra Leone.

138 Burning of charcoal piles; Czechoslovakia.

139 The influence of mechanized logging on a forest in Malaya: the numbers denote percentage of the base area of all trees more than 10 centimeters thick.

140

140 Man-made terraces for vineyards and fruit orchards; south Moravia, Czechoslovakia.

141 Rice fields in a clearing in a tropical rainforest; west Africa.

142 Herds of goats can cause the final destruction of the forest; southern Greece.

(Fig. 139). One great drawback is that only commercially established timbers are exported and of the many kinds available this represents only a limited range. For example, eighty per cent of the timber exported by Ghana comes only from the species: *Triplochiton scleroxylon*, *Entandrophragma cylindricum*, *E. utile* and *Mimusops heckelii*. In the tropics experienced tree-hunters roam through the forest looking for specimens of these species, which are then removed from the forest individually. In former times the felled trunks of these huge giants were hauled by hand to the nearest river and then floated down to collecting points. Nowadays a bulldozer clears the way to each such tree and the bole is loaded directly on to the truck in the forest. Such wasteful exploitation results in the reckless destruction of large tracts of forest and the squandering of wood which with today's improved mechanical and chemical methods could be put to good use. In all probability there are hundreds of species other than the ones demanded, presently growing in the forests of America, Africa and Indo-Malaysia and which could meet the most exacting demands as well as reduce the consumption of softwood in the countries of the temperate zone. Increased utilization of such timber resources could raise the national income of the developing nations while the forests of industrial regions could be used for water management and recreation.

In Europe, where forests include a far smaller range of species, experts long ago learned to exploit all the main kinds of coniferous and

141

142

143 A tree purposely burned when clearing ground for a field in the tropical forest; Ghana.

144 Fire, axe and saw—tools used by man to eradicate forest.

145 A burned Bosnian Pine *(Pinus leucodermis)* at an altitude of 2,000 metres; Olympic Mountains, Greece.

broadleaved trees. It was found that each species has different physical characteristics which can be used to advantage in various fields of industry. Spruce, pine, larch, oak, beech, maple and ash continue to be the most versatile woods. Industrial North America and Australia have a far greater range of woods at their disposal and have gradually learned to exploit them.

The demands of agriculture

That the world's forests are fast disappearing because of the demand for timber is a sad state of affairs, but a far greater cause for concern is the gradual recession of the forest brought about by the demands of agriculture and the rearing of livestock. In the countryside these two spheres vital to man's existence are often in opposition. Profitable agricultural products mostly require full sunlight and cannot be grown in the shade of the forest. Furthermore, most domestic animals feed by grazing on open pastures rather than foraging among forest trees and shrubs. To provide food for the world's vast populace the forest must

146 Robust forest trees finally end up as blocks suitable only for fuel.

147 Eradication of a flood plain forest on the site of a planned canal.

146

make way for arable land and pastures, rice fields (Fig. 141) and vineyards (Fig. 140).

Although it is possible for forests and agricultural land to exist side by side, history has shown that certain reasonable limits have to be respected. Centres of ancient civilizations in central China, India, Mesopotamia, on the Lower Nile and throughout the Mediterranean region were relentlessly devastated by the felling of forests to clear land for farming and by the uncontrolled grazing of cattle (Fig.142). This invariably led to the disruption of the water balance in the countryside, the drying out of the land, the loss of fertile soil and the general degradation of the environment.

The border between open country and the forest is not easily pushed back. The forest always has a certain stability and endurance and man unequipped with heavy machinery cannot easily conquer it. That is why, since time immemorial, man has used fire in his struggle with the forest. On the edge of pastures in Europe's mountains, as around clearings in the tropical rainforest, large trees marked by fire are always to be found (Fig. 143 and 145). Where rapid clearing of land is necessary huge areas are burned and felled (Fig. 144). With the use of heavy machinery such as caterpillar tractors, bulldozers, excavators, mobile cranes and efficient power saws even the most vigorous forest is soon turned into chips, logs and ashes (Fig. 146, 147 and 149).

The climate may also contribute to the de-

147

148

148 A giant dredge used to excavate sand undermines the foundations of a healthy forest.

149 Burning useless remnants of costly tropical woods; Ghana.

149

structive influence of fire. During the dry season fire is a great menace to the forest, particularly round the periphery of grassland steppes and subtropical savannas. For instance, in Africa it is estimated that as a result of man-made fires the area of closed savanna woodlands and semideciduous rainforests has decreased by as much as a half. Fires on mountainsides are particularly destructive, for the flames, drawn by the hot air currents, rapidly spread over vast areas.

The problem of pollution

The growth of industry during the second half of the twentieth century has brought about new and more drastic forms of destruction. Surface mining of various ores, coal, gravel sand, bauxite, kaolin and other raw materials (Fig. 148) requires the felling of large tracts of forest and causes irrevocable changes in the structure, water balance and surface climate of the countryside. The return of the forest to such devastated land is very difficult if not impossible. Far and wide round industrial centres, large power plants and conurbations, the remaining forests are dying as a result of atmospheric pollution by gaseous and solid particles spewed into the air by chimneys. After a number of years the action of these poisons changes the composition of the forests. Various trees and shrubs are killed sooner or later, depending on their degree of resistance, and the fertility of the habitat is generally disrupted (Fig. 150, 151 and 153). These changes are clearly evident in the neighbourhood of such centres of pollution and can thus be evaluated from the economic as well as the environmental viewpoint. The fact that all of central Europe is already exposed to increased concentrations of sulphur dioxide and other poisonous products in the atmosphere is a warning to other parts of the world.

Tropical forests

The destruction of tropical forests is also proceeding at a rapid pace. According to FAO (Food and Agriculture Organization of the United Nations), some forty per cent of all tropical forests have been destroyed during the past 150 years. The population of tropical countries is increasing by geometric progression and thus it is necessary to assume that by the year 2,000 all tropical forests will have been affected in one way or another. Either they will have been cut down and the land used for agriculture or urban development, or else they

150

150 A ruined forest in the vicinity of a strong source of pollution.

151 A dying forest exposed for years to the effects of industrial pollution

151

152 A plantation of Norway Spruce *(Picea abies)* in hilly country.

152

153

153 Dead trees on a slope exposed to the effects of harmful waste gases.

will have been transformed into semiwild bush consisting of several weedy species of trees, lianas and shrubs. With the low level of industrial development and the antiquated methods of farming used in these areas it must be expected that by the end of the century some 500 million people will obtain food by extensive exploitation of the tropical forest—by the primitive cultivation of crops on ground cleared by burning.

This process, which is almost impossible to check, will lead to the destruction of a great part of the world's flora and fauna, for it is well known that tropical forests contain a great number of plant and animal species that have evolved during the course of time and have not yet been described. Each of these hitherto unkown plant and animal species is a phenomenon in its own right. It has emerged during a process of evolution lasting many millions of years, it has its place in the realm of nature, and it is doubtless also of potential use to mankind. It would seem, then, that the scientific investigation of tropical forests may well be more urgently needed than space research or flights to the moon.

The re-establishment of forests

In a congenial environment, where climate and soil are conducive to its growth, the forest

154

154 A planted grove of Silver Birch *(Betula pendula)*.

155 Alders planted on former pastureland, interspersed with naturally regenerated spruce.

155

156

156 Artificial park-like mixed forest; Harriman State Park, New York.

157 Pastureland forest in Bulgaria.

158 An old olive grove in Montenegro, Yugoslavia.

157

158

need not be permanently and irrevocably on the retreat. When man ceases his stubborn acts of deforestation, by moving out of a territory for example, or changing his method of using the land, the forest re-establishes itself. Such returns occurred after the Thirty Years' War, the Second World War, and even more recently when the import of cheap grain struck a paralyzing blow to agriculture in less fertile regions. Well-documented is the reoccupation of land by forests in North America, where mobility is a common characteristic of both people and industry. In the second half of the nineteenth century, when the fields of New England were abandoned by farmers in favour of the fertile prairies of the Middle West, the Weymouth Pine *(Pinus strobus)* spontaneously reoccupied the land, forming forests which by the beginning of the twentieth century were already of exploitable dimensions and gave rise to a large timber industry. The original forests were replaced by mixed forests very similar in composition. In the vast tracts of closed forest the stone fences that once marked farm boundaries are the only traces of what was once farmland.

In tropical regions the reoccupation of cleared land by forest is even more dramatic. Many abandoned cities were rapidly covered by lush jungle growth. Growing on the masonry of old cities and monasteries in Central America and south-east Asia are giant trees and roots that twist and wind through the ruins. The secondary forest that develops on the site of the original rainforest naturally never resembles it in structure and composition of species. Absent, first of all, are the original giants of the top layer that sprout with difficulty amidst the dense growth of the pioneer trees. Many species of trees as well as animals return to the site of their natural occurrence after hundreds of years.

Forest also reoccupies land in Africa, where the inhabitants who cleared it have either died or moved away because of epidemics of parasitic diseases or wars. After several hundreds of years there are only negligible traces of the villages that once stood in such places and the tall forest looks just as vigorous as those on land that was never cleared by man. Detailed analysis, however, always reveals that certain components of the original forest are absent. The most notable absentees are the various species of forest mammals which require time and land connections with the centres of distribution of each species.

Thus far, man has been considered as a casual ecological factor affecting the forest cover of the globe—and mostly in the negative sense. It must be acknowledged, however, that today the extensive changes in the world's forests are being made with the active and sometimes planned participation of specialists—foresters, sylviculturists, and arborists. Their participation in the transformation and exploitation of forest is considered in

Chapter 8, but a few general comments may be appropriate here.

Forests throughout the world are changing not only in the area they cover, but first and foremost in composition of species and general structure. In developed countries all forested land is managed by foresters, who change the composition of the forests by planned cultivation and cutting. Forests are also affected by many other factors—mainly agriculture, water management, urbanization, landscape architecture, and nature conservation. Other decisive factors in the care of forests are the social system and the system of land ownership in a particular country.

Planned planting, thinning and cutting in developed regions is producing forests that are a far cry from the original natural forests. For example quite common nowadays in the hilly country of central Europe are artificial stands of Norway spruce, which began to be spread by the Saxon School of Forestry more than 250 years ago (Fig. 152). In forests around cities such new forms as pure groves of the silver birch *(Betula pendula)* are found (Fig. 154). In the middle of large forest complexes remarkable combinations of leaf-trees and conifers can be found which on closer investigation prove to be the initial stage of forestation on what were formerly meadows (Fig. 155). Open woodlands on fertile soils (Fig. 156) are either remnants left by spot felling or, alternatively, planned recreational woodlands. Forests that have developed on land previously used for grazing always have a distinctive character (Fig. 157). Some abandoned pastures turn into open groves (Fig. 158). Quite bizarre in appearance are the remains of poplar and willow flood plain forests which for decades have served man as a source of material for wickerwork (Fig. 160).

Today, as a result of modern organization and communication with all parts of the world, foresters can import seeds or cuttings of forest trees from any other country. If they find the rapid-growing Grand Fir *(Abies grandis)* or balsam poplar best suited to their purpose, they have them shipped from North America.

Conversely, when the vigorous Scots pine attracted the notice of American foresters they planted large stands of this important European timber tree in the United States.

The introduction of exotic trees and shrubs to local forests changed the appearance of entire sections of the landscape in many parts of the world. Exotic trees and shrubs were originally introduced into parks and gardens purely for ornamental purposes but forestry soon discovered the great potential of such trees both as a source of rapidly growing timber, often with improved qualities, and as an instrument in the amelioration of devastated areas.

An important example is the international spread of the black locust *(Robinia pseudoacacia)* in Europe (Fig. 159). This tree of the legume family is native to North America

159 The Black Locust *(Robinia pseudoacacia)* has become established in Europe.

160 Pollarded White Willow *(Salix alba)*.

159

161

161 A plantation of fast-growing poplars.

162 An oak plantation with trees in regular rows in central Bohemian hilly country; Czechoslovakia.

163 Stumps and disorderly stands of *Populus pyramidalis* in Macedonia, Yugoslavia.

162

163

164 A huge scree cone (intersected by a stream), formed by water erosion as a result of deforestation.

165 Much deforested but inviting landscape in the Italian Dolomites.

165

164

where it grows at lower elevations in the Appalachians and in the mountains west of the middle reaches of the Mississipi. In its native land it is a common inconspicuous tree of little worth, scattered throughout the rich mixed forests together with such other common members of the legume family as the Honeylocust *(Gleditsia triacanthos)* and Kentucky Coffeetree *(Gymnocladus dioicus)*.

The black locust arrived in Europe via Germany as far back as the early seventeenth century. At first it was grown as an ornamental tree, but later it began to be prized as food for bees, as a source of durable poles for vineyards and above all as a tree particularly effective in the binding of dry sands and deforested slopes. It thrived particularly in the warmer regions of Europe where it began to spread very rapidly. Its successful naturalization and invasion of native forests is the result of its great fruitfulness, rapid spread by means of root suckers, and extraordinary competitive power.

The spread of the black locust greatly changed the character of forests on poor sandy soils and sunny hillsides. Other species are hard put to survive in the company of this tree, and its marked effect on the soil chemistry (its roots exude a substance that affects all plants and animals) depletes the flora and fauna of entire regions. In Europe similar vigour is exhibited by certain other trees such as the Box Elder *(Acer negundo)* and Northern Red Oak *(Quercus rubra)*.

Introduced trees have been known to spread widely in other parts of the world as well. One such example is the Neem *(Azadirachta indica)* which is native to India. An introduction to Africa's savannas, its resistance to drought has it winning out over most of the native trees, which are gradually being ousted by its spread. The same is true of many exotic species of the genus *Cassia* and *Acacia*.

The introduction of new trees need not always be from one continent to another. The forests of central Europe, for instance, are full of the European larch, which, before it began to be planted on a wider scale some two hundred years ago, occurred only in the mountain regions of the Alps, Sudetens and Carpathians. Changes in the composition of forests are sometimes also brought about by the introduction of herbaceous plants. In the flood plain forests alongside the large rivers of central Europe, such plants as Golden Rod *(Solidago gigantea)*, Michaelmas Daisy *(Aster novibelgii)* and Policeman's Helmet *(Impatiens roylei)* are spreading and ousting all native flora. The ground flora in these forests is changing into a uniform carpet of flowers that must needs lead to uneven exploitation of the soil and decreased diversity of the ecosystem.

Cultivated stands usually have the look of an orthodox plantation with the trees spaced regularly in rows (Fig. 161 and 162). Though such a regular conformation is expedient from the economical, technological and organizational viewpoint, it must be acknowledged that such precise geometrical patterns are foreign to the forest ecosystem.

Sometimes it is difficult to strike the right balance and decide how much forest man needs in his environment. Parklands composed of conifers interlaced with alpine turf (Fig. 165) may be advantageous from the economic viewpoint as well as from the viewpoint of man's well-being. There can be no doubts, however, when forest land is threatened by erosion, or where only thorny thickets and herbaceous weeds follow in the forest's wake (Fig. 164). Nowhere should the forest's fate be put in jeopardy, for the outcome would be the tragic fate that befell the once-wooded countries of the Middle East and the Mediterranean region.

166 Pines whipped by winds on the Porquerolles Islands off the southern coast of France.

THE INFLUENCE OF THE ENVIRONMENT

In the broader geographical context, climate is the decisive factor that determines the composition of the forest, temperature and annual rainfall being the most significant factors (Fig. 167). However, on a smaller scale the characteristics of a forest change from place to place according to the topography, soil conditions and distribution of water. Forest trees as well as other forest plants and all forest animals are influenced, directly or indirectly, by the physical and chemical properties of the geologic substrate, the configuration and inclination of slopes as well as their orientation, the depth, richness and moisture content of the soil, and the atmosphere near the ground. These factors affect the appearance, growth and development of the entire forest (Fig. 168). The forest, in turn, modifies certain primary characteristics of the landscape, as is readily demonstrated when such a forest is destroyed by some natural catastrophe or is cut down by man. The forest does not play merely a passive role, but actively participates in shaping the living environment. The life processes and activities of the forest's inhabitants alter many characteristics of the soil and climate near the ground, and sometimes even the surface appearance of the landscape.

The effect of relief

The principal factor which accounts for the different types of forest is the configuration of the earth's surface—its relief. Hills, mountains, ridges, valleys, terraces, plateaus and faults form the background against which the life of the forest is set. The angle at which a particular slope is inclined, the direction in which it faces, and also its height above sea-level all combine to determine the supply of heat, light, water and nourishment. Only rarely are completely level landscapes encountered. Forest territory is usually divided into varying degrees of high and low ground by a network of watercourses, some of which may be shallow and others very deep.

In land lying at the foot of a mountain chain, and in the mountains themselves, the diversity of the forest cover is very striking. The slope of rock strata and the orientation of the slopes cause great differences in the composition and quality of the forest (Fig. 169). Quite different from the rest are peaks and jutting ridges which usually have a shallow layer of stony soil, insufficient water for plant roots and a drier and windier climate. In such an environment plants suffer from lack of water and nutrients and are damaged by gales and glaze ice—produced when rain falls on vegetation which is at a subzero temperature. This summit phenomenon (Fig. 170 and 172) is characterized by stunted and crooked growth and a more open forest canopy than is found on lower slopes and valley bottoms (Fig. 171 and 173). In Europe and North America summit forests are deformed particularly by the very heavy ice cover that forms when rain falls on greatly undercooled branches or needles.

In damp tropical jungles, forests sometimes become less vigorous on the hilltops and give way to open or savanna-like woodlands. This is often caused by the fact that hardened laterite or bauxite substrates, which are inadequate for the nourishment of the forest, are near the surface of the soil on such summits. In higher tropical mountains that reach to the cloud level, however, the atmospheric moisture may increase on the summits (Fig. 174) and this is immediately evidenced by the great abundance of mosses, ferns and orchids growing in the crowns of trees.

The summit phenomenon also manifests itself in the composition of the animal community. Unlike the flora, which is usually impoverished in elevated places, there is a greater number of insect and vertebrate species. According to some entomologists it is possible to capture on such a summit all the species which occur in the surrounding slopes of the forest.

In low-lying areas and at the bottom of valleys the forest environment, and therefore also the forest cover, is markedly different from that found on the summits (Fig. 175). Water flows to the base of the slopes and carries down quantities of dissolved nutrients, mineral particles and humus. In general the soil in depressions is deeper, springs often gush forth there, and the water table is close to the surface. This accumulation of soil and the reliable supply of water are exploited to the full by forest trees and all forest creatures

167

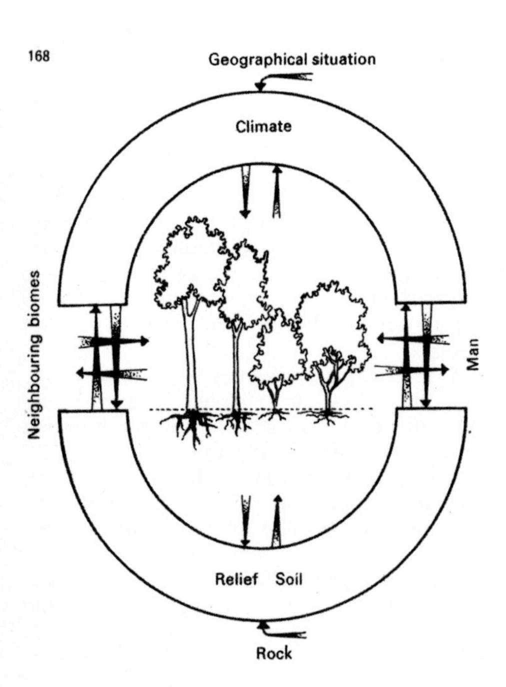

167 The life of the forest is constantly affected by the powerful forces of the environment.

168 General diagram showing how the forest is influenced by the environment.

living on or in the soil. That is why it is at the foot of slopes that the largest and tallest trees and likewise a wealth of tall herbaceous plants, such as various species of *Aconitum*, *Aruncus*, the Umbelliferae family and ferns are found (Fig. 177).

However, in low lying areas and valley bottoms where water does not run off readily and the soil remains saturated for long periods the forest is usually poorly developed in terms of both growth and range of species. Waterlogged soil is deficient in oxygen and may contain toxic concentrations of carbon dioxide, so the roots of common forest trees and herbs cannot respire. The only trees that grow suc-

169 The influence of the slope of rock strata on the height and vigour of the forest: on a slope where the strata lie at right angles to the surface the forest roots better and is most vigorous.

170 Only open woodland made up of crooked oaks *(Quercus)* and pines *(Pinus)* grows on the ridge.

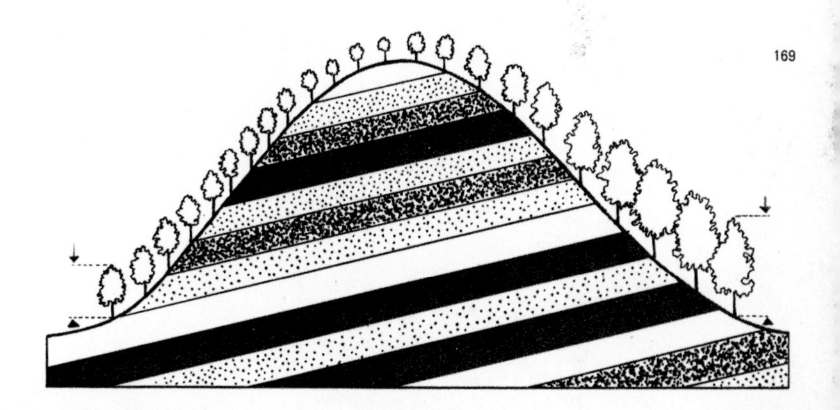

169

170

171

171 Only the undemanding Scots Pine *(Pinus sylvestris)* will grow on such a rocky summit.

172

cessfully in swamps are specially adapted alders, willows, poplars, birches and pines, which in places that are permanently waterlogged give way to the better adapted grasses, sedges, and mosses. In tropical regions palms, bamboos and pandanus grow in such places as do grasses and certain mosses and some large, broadleaved species of the Marantaceae and Zingiberaceae families.

At the base of deeply cut valleys and ravines such as are often found in sandstone, dolomite and limestone regions (Fig. 176) the forest is affected also by long periods of, or perpetual shading from the sun. Direct sunlight penetrates with difficulty and for only a short time each day, so the forest must make do mostly with diffused light. Furthermore, insufficient illumination is usually accompanied by low temperatures for diffused light has little energy, and cold air flows downwards and collects at the bottom of narrow ravines. In such cold, shaded places a type of forest may be found that is quite foreign to the particular region, being more closely related to mountain forests or those found in the north.

On middle slopes located between the summit and the base there are various types of forest, depending on the orientation of the slope, the depth of fertile soil and the moisture content (Fig. 182). On slopes with an angle of

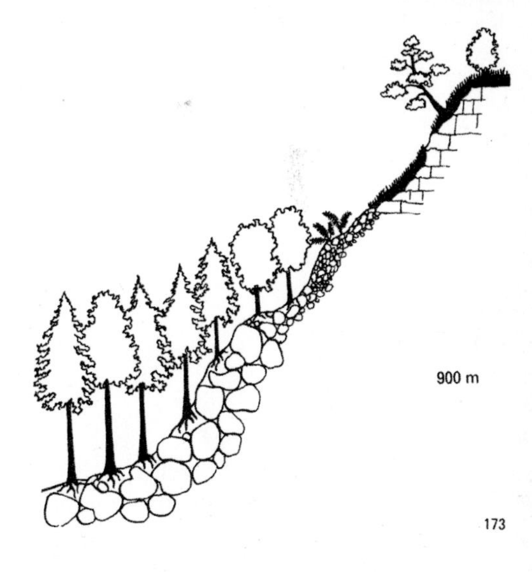

173

172 The height of mountain coniferous forest varies with its location on the slopes.

173 Forest and non-forest ecosystem in the area of a rocky peak; the shallow layer of soil at the top produces only low, stunted trees, whereas the deeper soil layer lower down the slope allows for a dense, tall forest.

174 The composition of the forest changes even on the summits of low ridges in the heart of the tropical jungle; mist forest in Ecuador.

175 Forest swamps form in low-lying areas and these are not conducive to the growth of demanding forest trees.

174

30° to 40° there is a thick layer of soil largely held in place by the roots of trees and herbaceous plants. On very steep slopes, water washes away fine earth and humus thus bringing the underlying rock closer to the surface. In such shallow soil a forest has difficulty in retaining a foothold (Fig. 178 and 179), though some pioneering trees may sprout in rock crevices and on small ledges formed by uneven weathering of the rock. The roots of forest trees help to hold the soil on steep slopes, but roots growing in rock crevices haste rock disintegration and weathering. Furthermore, soil falls under its own weight, and loosened stones may batter tree trunks lower down the slope (Fig. 180, 181 and 183). The stability of the forest on steep slopes may also be disrupted by large scale catastrophes. Soil, boulders and vegetation may be swept away by avalanches while heavy rain may make the soil waterlogged, causing it to slide, along with the forest, down the slope into the valley (Fig. 184). Tracts left bare by landslides can be found in mountains throughout the world. The return of the forest to such areas usually takes several hundred years.

175

176 This deep ravine in a limestone region has comparatively more light in winter than in summer when the trees are in full leaf.

177 The influence of an undulating ground surface on the height of a tropical forest; trees growing in the valley are much taller than trees growing on elevated areas.

176

From hilltop to valley, then, there exists a chain of various types of forest on different qualities of soil. On the upper slopes it starts with low forest that obtains nourishment only from a weathering rock substrate impoverished by erosion. Lower down, on the middle slopes, the forest takes nourishment both from the soil in situ and from that washed down from above. At the foot of the slope the soil is enriched by nutrients and water from all of the hillside above. This progression in the quality of the soil has a pronounced effect on tree growth both in height and girth. For example, in the mountains of central Europe spruces growing in a valley may reach a height of thirty metres whereas 200 to 300 metres higher, near the summit, they are only ten metres tall. When trees are removed in the temperate and tropical zones the bared forest margin reveals that trees in depressions are always taller than those on raised ground, so any unevenness of the terrain may be entirely concealed by the level of the tree tops.

The effect of water

Water plays an important role in the life of the forest and the forest in turn markedly affects the water balance in the area. Water in all its forms—vapour, liquid and ice—has a considerable influence on the range and coexistence of forest plants and animals. Particularly dramatic is the forest's encounter with running water at the bottom of the valleys (Fig. 185 and 186). Along the sides of streams and rivers there are always whole series of shrubs

177

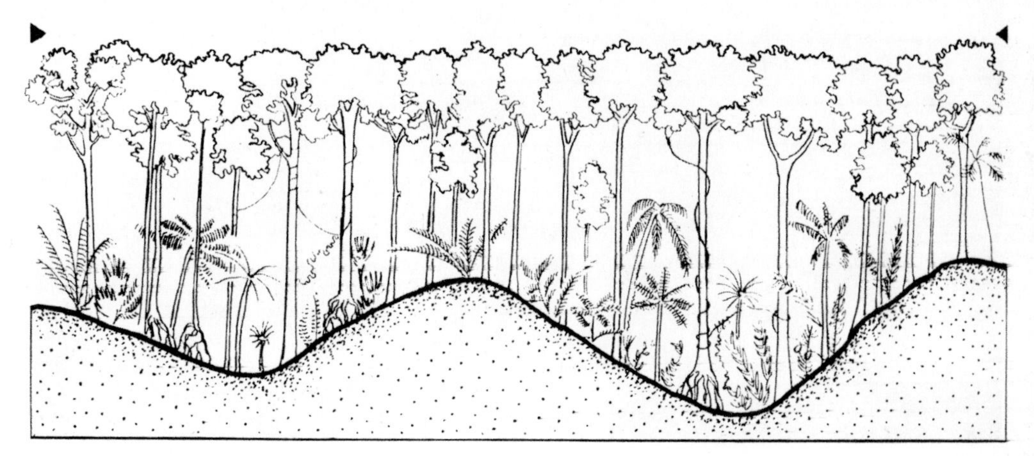

178 Sandstone which weathers rapidly is practically without vegetation except for Silver Birch *(Betula pendula)* and Scots Pine *(Pinus sylvestris).*

and trees (Fig. 188). Factors important to their existence are the mechanical force of the current, the soil this current deposits or washes away, and the temperature of the water and the ice that rushes down during the spring thaw. Usually growing near the edge of the water and on exposed islands in the middle of the river are shrubby woodlands composed mostly in the temperate regions of the northern hemisphere of various willows, dogwoods *(Cornus)* and tamarisks *(Tamarix).*

Exposed mountains of the High Tatras affect air currents, rainfall and temperature.

In central Europe southern slopes with a gradient of 30°, where forest gives way to grassy steppe, are the warmest.

Pioneer trees at the polar limit in Alaska are the Black Spruce *(Picea mariana)* and the White Spruce *(Picea glauca)*.

On the banks of the Amazon River the water rises several metres during the rainy season.

Soil is both deposited and carried away by the winding river in the flood plain, making for a heterogenous forest.

These shrubs are able to withstand the mechanical damage caused by the water current and ice floes, and they multiply readily from seed which is produced in abundance as well as by the rooting of twigs that have broken off. Behind this buffer of waterside shrubs there is a belt of trees including willows, poplars, birches, alders and maples together with certain conifers which prefer an open aspect. These pioneer trees and their companion herbs stand up well to the high water table and also tolerate the frequent floods which occur during the rainy season or following the spring thaw when the snow melts in the mountains (Fig. 187). In the temperate zone

179 It is a miracle that the Arolla Pine *(Pinus cembra)* maintains a foothold on this steep granite rock in the High Tatras.

179

of North America and Eurasia the number of such pioneer plants is small and even in the tropics the variety of plant species is not very large in waterside woods or severely flooded regions. In the tropics such forests include, pandanus, bamboo and palm trees (Fig. 189 and 190), all of which survive even long-term and repeated flooding. Many tropical trees are well equipped for life in flood-plains. They readily grow adventitious roots which support partially uprooted trees, and can produce various types of modified roots that facilitate respiration in waterlogged soil which contains an insufficient supply of oxygen. Similar characteristics are exhibited by the baldcypress in the subtropical regions of America and in the temperate zone by various alders.

Rivers

The forests in the flood plains farther from the river channel are extremely vigorous. They are very productive and possess a great diversity of species which utilize the rich alluvial soil. In Europe it is in these flood plain forests that the largest oak, ash, elm, maple and spruce trees are found. In the damp tropics the largest giants grow on loamy alluvium. Fertile silt, which improves the quality of the soil, is carried to the flood plain forests by occasional floods (Fig. 191 and 192). If they are of short duration, such floods do not destroy the forest vegetation but they may force animals to move to drier ground for a time.

The soil in flood plains is the result of the action of water over thousands of years. The changing force of the stream of water builds up alternating layers of sand, gravel, loam and clay in flood plain forests, resulting in differences in the types of forests found there. Flowing water is also important for the dispersal of plant seeds and in the migration of animals

180 Mountain slopes where a thick blanket of snow and avalanches are a yearly occurrence are covered with shrubby stands of broadleaf trees.

181 Common Beeches *(Fagus sylvatica)* on a steep slope in the Malá Fatra Mts. have sabre-shaped boles.

180

181

182 Section through a valley with colder north-facing slope on the left and warmer south-facing slope on the right; top—natural distribution of various types of forest; bottom—remnants of forest and substitute ecosystems created by the farmer.

183 Trunk of a Common Beech *(Fagus sylvatica)* battered by falling stones.

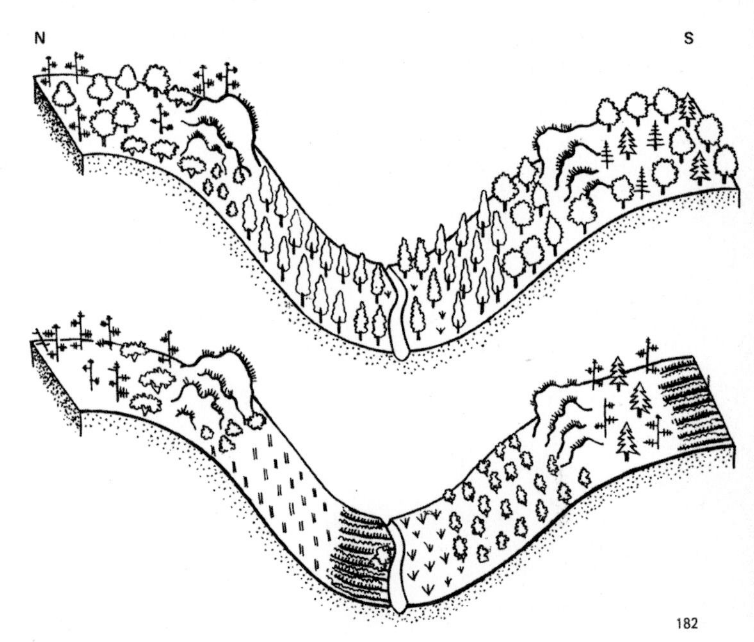

adapted to water. In the lowlands at the foot of high mountains, flood plain forests and waterside woodlands often include mountain species the seeds of which have been carried down by the stream. Frequent examples of such settlers in the forests of North America and Europe are mountain species of spruce, pine and fir (Fig. 193). The coexistence of river and forest is no peaceful affair and the shoreline continually witnesses the drama of creation and destruction. As the stream meanders through the flood plain it washes away the banks and undermines the roots of the fully grown trees. Sooner or later even the strongest trees become uprooted (Fig. 194 and 196) and then the fallen trees become the cause of further changes. New patterns of still water and eddies are formed, the course of the stream is changed, and soil is eroded and deposited in other places. Even such firmly rooted trees as oak, hickory, elm and ash are affected by the lateral washing away of the soil. The European Alder *(Alnus glutinosa)* is unusual in that it is particularly effective in holding the partially eroded banks of Europe's streams and rivers.

183

Seas

Forest, however, does not tolerate a permanent aquatic environment. There is no such thing as an aquatic tree, and the whole life

184 Landslides leave scars on forested slopes; Cuchillas de Toa range, eastern Cuba.

185 Zonality of woodland on the banks of a river: shrub willows by the edge, hardwoods higher up.

186 Alders, ashes and maples thrive alongside streams near Prague, Czechoslovakia.

184

185

186

form of woody plants is adapted for a terrestrial existence. Sometimes trees appear to be growing out of water but this is only a temporary phase. In mangrove woodlands the trees stand deep in water twice a day at high tide and during the spring tides large areas may appear to be growing in open sea (Fig. 195). High tide, however, is always followed by ebb tide and it is this that provides conditions suitable for the spread and germination of the specialized species of trees growing in mangrove woodlands (Fig. 197). In other instances the aquatic and terrestrial phases in the life of a forest alternate during the year.

Mangrove woodlands are quite the most peculiar forests in the world. They occur on the seacoast in the tropical zone reaching the latitude of 32° N on the northern hemisphere (Bermudaa southern Japan) and the extreme of 40° S on the southern hemisphere (South Africa, Australia and New Zealand). 30° N and 40° S which encompass Bermuda, southern Japan, southern Africa, Australia and New Zealand. Falls in temperature to below freezing point limit the spread of this type of forest in the temperate zone because none of the species that grow there tolerate frost. Man-

187

187 Flooded forest after a drop in temperature; the rate of thawing of snow and ice is affected by the temperature of the treetrunks.

188 Zonality of shrub and forest on the banks of a mountain river: 1—river, 2—grasses and sedges, 3—initial stage of scrub growth, 4—thickets of shrub willow, 5—mixed coniferous forest.

189 Dense thickets of broadleaved species and palms beside a tropical river, Ghana.

188

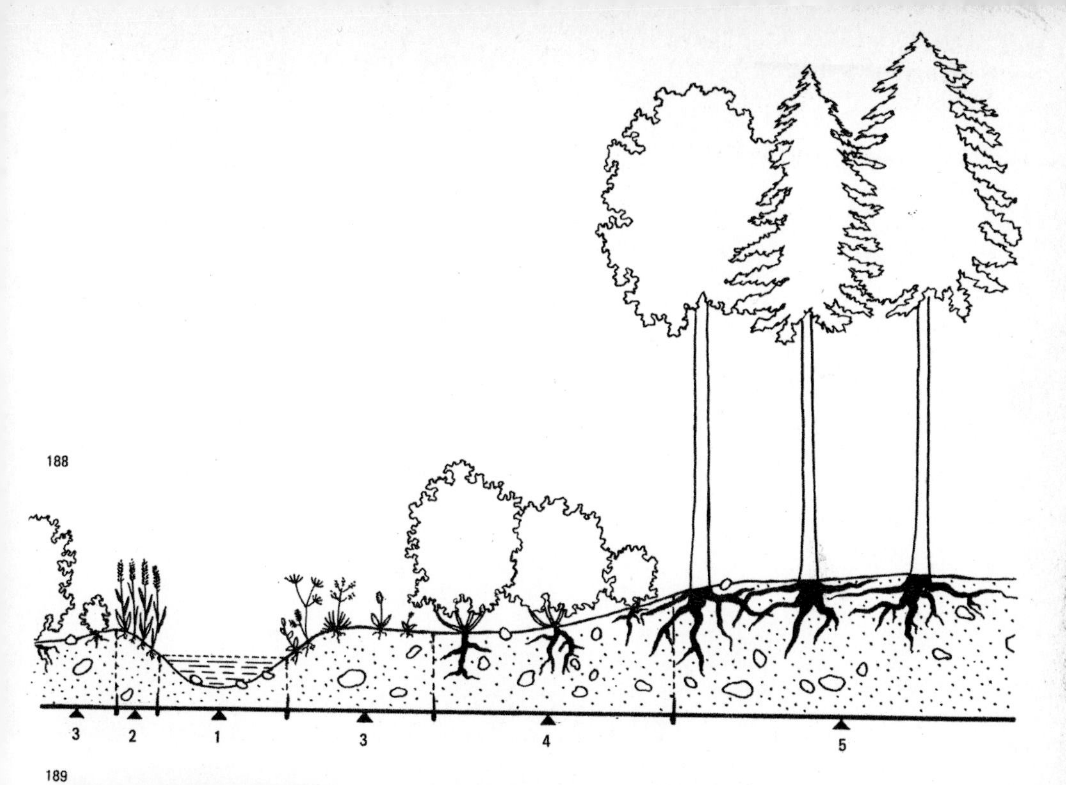

189

190

grove woodlands throughout the world include only about thirty species divided amongst some ten genera, namely *Rhizophora, Bruguiera, Ceriops, Sonneratia, Xylocarpus, Lumnitzera, Aegiceras, Avicennia, Laguncularia* and *Conocarpus*. The species are unequally distributed, more than two-thirds growing on the shores of the Indian and Pacific Oceans, and the remainder on the shores of the Atlantic. The first region is also the home of the palm tree *Nipa fruticans* which grows on large tracts side by side with the broad-leaved mangrove stands.

Many species of trees in mangrove woodlands grow to a height of twenty metres. Most have highly modified roots, adapted so that they are particularly suited to the changing level of the sea, tidal ebb and flow and the saturated condition of the soil (Fig. 198). Mangroves are halophytic species and their leaves are morphologically and physiologically adapted to high concentrations of salt in the soil. They are leathery and able effectively to limit transpiration. Even with minimum transpiration the leaves become charged with salt which some species of the genus *Avicennia,* for instance, eliminate in the form of crystals on the underside of the leaves. The way in which the trees of mangrove woodlands protect themselves against excessive concentrations of salt in the leaves has not yet been fully investigated. Nor has a satisfactory explanation been found for the fact that young seedlings which grow attached to the parent plant contain practically no traces of salt, or for the fact that their tissues and water regime begin to resemble those of adult specimens only after they have rooted in the soil.

The growth of seedlings in the tops of the trees is another adaptation to tidal ebb and flow and the character of the muddy shore, for the seeds and light fruits would be easily car-

190 *Raphia hookeri* palms cover waterlogged soil around tropical rivers.

191 Of the European trees the Common Alder *(Alnus glutinosa)* is best able to tolerate permanently waterlogged soil.

191

ried away by seawater if they fell to the ground. When a well-developed seedling with a heavy, pointed, primary root falls to the ground, it does so in such a way as to bury the root in the mud and thus may remain in place despite the tidal flow. Successful rooting and permanent establishment of a new tree may be further aided if it falls during the period of the neap when the tides are lowest, and the difference between high and low tides is smallest, for at this time the swamps are inundated to a lesser degree and for a shorter time.

Respiration in submerged roots in mangrove swamps is made possible by special breathing roots. In members of the genus *Avicennia* these resemble slender pegs growing out of the ground, in members of the genus *Sonneratia* they are thick conical structures (Fig. 200), and in members of the genus *Bruguiera* and *Ceriops* they are knee-like organs 199). Breathing roots have special pores or lenticels that permit air to enter while keeping seawater out. At high tide when the mangrove woodland is inundated and the entire root system is submerged, use is made of the oxygen stored in the many intercellular spaces of the roots for respiration. This causes reduced pressure in the root tissues which is equalized only when the roots are again above water and can absorb oxygen from the air through the lenticels.

Various species of trees growing in mangrove swamps differ not only in structure and in their physiological functions but also in their distribution along the shoreline. Growing farthest out in the sea are members of the genus *Sonneratia,* closer to shore grows a belt of *Rhizophora,* succeeded by a zone of *Ceriops* and *Bruguiera,* and closest to the land are

192

members of the genus *Avicennia.* Besides specialized trees, mangrove woodlands are associated with specialized flora and fauna. All mangrove woodlands, for example, contain the fern *Acrostichum aureum,* and all have typical populations of crabs and the curious fishes of the genus *Periophthalmus* that 'hop' about on the roots and trunks of the trees.

Springs

The mosaic of forests throughout the world evidences the influence of springs rising to the surface from the underlying rock at various levels and the foot of slopes. A spring may rise in a particular place into a hollow where the water forms a small pool. Alternatively it may rise at an angle and runs down the slope in the form of a stream. Sometimes a spring percolates through a layer of soil and vegetation

192 A long-term flood formed rings of decaying algae on the trunks of a flood plain oak forest.

193 The Norway Spruce *(Picea abies)* spreads down from the mountains along the streams.

193

194

over a large area on a slope and only at the bottom of the valley does the water collect to form a surface stream. In this third instance, in the middle of a forest typical of the area, a small island of entirely different forest is found, having a ground flora dominated by various mosses and liverworts, and sometimes rushes and grasses. Only certain trees tolerate such wet soil. In Europe alders, willows, birches, and ashes, can do so and they are often found in association with spruce.

The development of forests round springs is characterized by two things—the trees usually have a rapid rate of growth with a shorter life span, and individual trees are more prone to being blown over. These processes result either in thickets of vigorously growing young trees or loose more mature woods with gaps left by uprooted trees.

Other effects of water

Water courses and lakes also affect the composition of forests by their influence on the humidity and temperature of the atmosphere. Sometimes evaporation from the surface of these bodies of water adds a great deal of water vapour to the air causing mists and heavy condensation when the atmosphere cools at night. Adjacent to rapids and waterfalls in tropical regions, (Fig. 202) more cold-resistant (mountain) types of forest can be found because the sprayed droplets of water evaporate in the air and on the surface of the vegetation, cooling the whole forest ecosystem to a level several degrees below that normal for the area. For this reason it is also possible to find, even at low altitudes in the middle of tropical jungle, plants and animals which are normally associated with high mountains. The reverse effect is obtained in areas where the calm surface of a river reflects the sun's rays on to neighbouring slopes thus contributing to the growth of forests which are normally associated with a much warmer climate. In the deeply cut valleys of central European rivers, such thermophilous forests are often replaced by vineyards.

The effect of the forest on water

Forests have a marked influence on the quan-

196

194 An English Oak *(Quercus robur)* with soil washed away from the roots on one side by the river.

195 Last islet of forest: *Sonneratia alba* on the coast of the Indian Ocean at high tide.

196 An uprooted tree causes changes in the flow of the river and effects the washing away and deposition of soil.

195

tity, annual fluctuation and quality of water in their vicinity. The strength and quality of springs (Fig. 201) depends in great measure on the composition, density and rooting ability of the forest trees. Long-term investigations have shown that forest cover regulates the level of water in streams and rivers, for forest soils readily absorb water and prevent a sudden rise of the water level. The steady percolation of water through the soil similarly reduces the likelihood of streams drying up. Forests also help to keep water clear of soil particles and bacterial impurities. Where mountain forests reach the cloud layer, they increase precipitation in that droplets of water from clouds and mist accumulate on the branches, leaves and needles, whence they fall to the ground and augment the soil moisture. If the soil is shaded this also benefits the water balance of the area because less water evaporates from the soil. On the other hand, the forest can reduce the amount of rain that reaches the soil for part of the precipitation is caught in the treetops and is drawn back into the air before it can reach the plant roots.

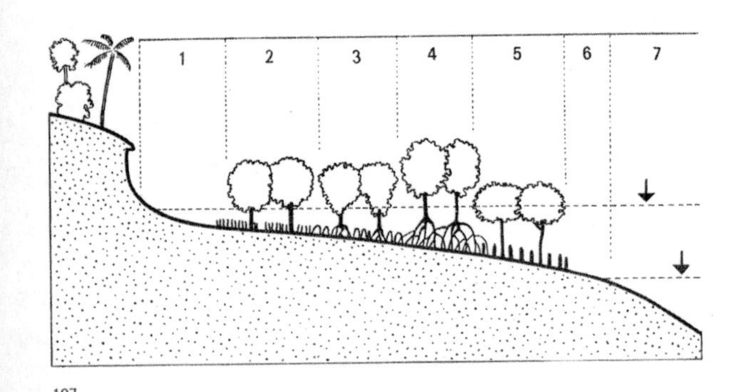

197

197 Zonality of mangrove woodlands on the coast of East Africa: 1 — barren, sand zone, 2 — *Avicennia maritima*, 3 — *Ceriops tagal*, 4 — *Rhizophora mucronata*, 5 — *Sonneratia alba*, 6 — zone of algae, 7 — sea with marks showing the extreme high tide and low tide levels.

198 *Rhizophora mucronata* of mangrove woodlands has good powers of regeneration.

198

199

The effect of light

The atmosphere in a forest is very different from that in open country (Fig. 206 and 207). During the day a great part of the solar radiation is caught by the treetops, and direct sunlight reaches the soil surface only through gaps in the canopy. Thus the level at which the greatest amount of radiant energy is received and also reflected back—and therefore the greatest amount of heat is lost—is transferred in forested areas to a level high above the ground. This is why the greatest daytime and night-time temperature fluctuations occur in the top layer of the canopy. Herbs growing on the forest floor receive less than one per cent of the light that falls on the treetops. Dimness within the forest is thus a typical characteristic of this environment. Through the gaps in the canopy, however, light penetrates to this dim region in the form of constantly moving flecks of light that follow the path of the sun through the heavens.

The amount of light that diffuses to the level of the trunks and to the surface of the soil in a forest depends on many factors. Most important are the morphological characteristics of the dominant species, in other words the

199 Knee-like roots round the trunk of *Ceriops tagal* in a mangrove woodland; Tanzania.

200 Breathing roots of *Sonneratia alba*.

200

201

202

201 The amount of water yielded by springs throughout the year is influenced by the forest.

202 The evaporation of water droplets always lowers the temperature around the waterfalls in the tropical forest; Malaya.

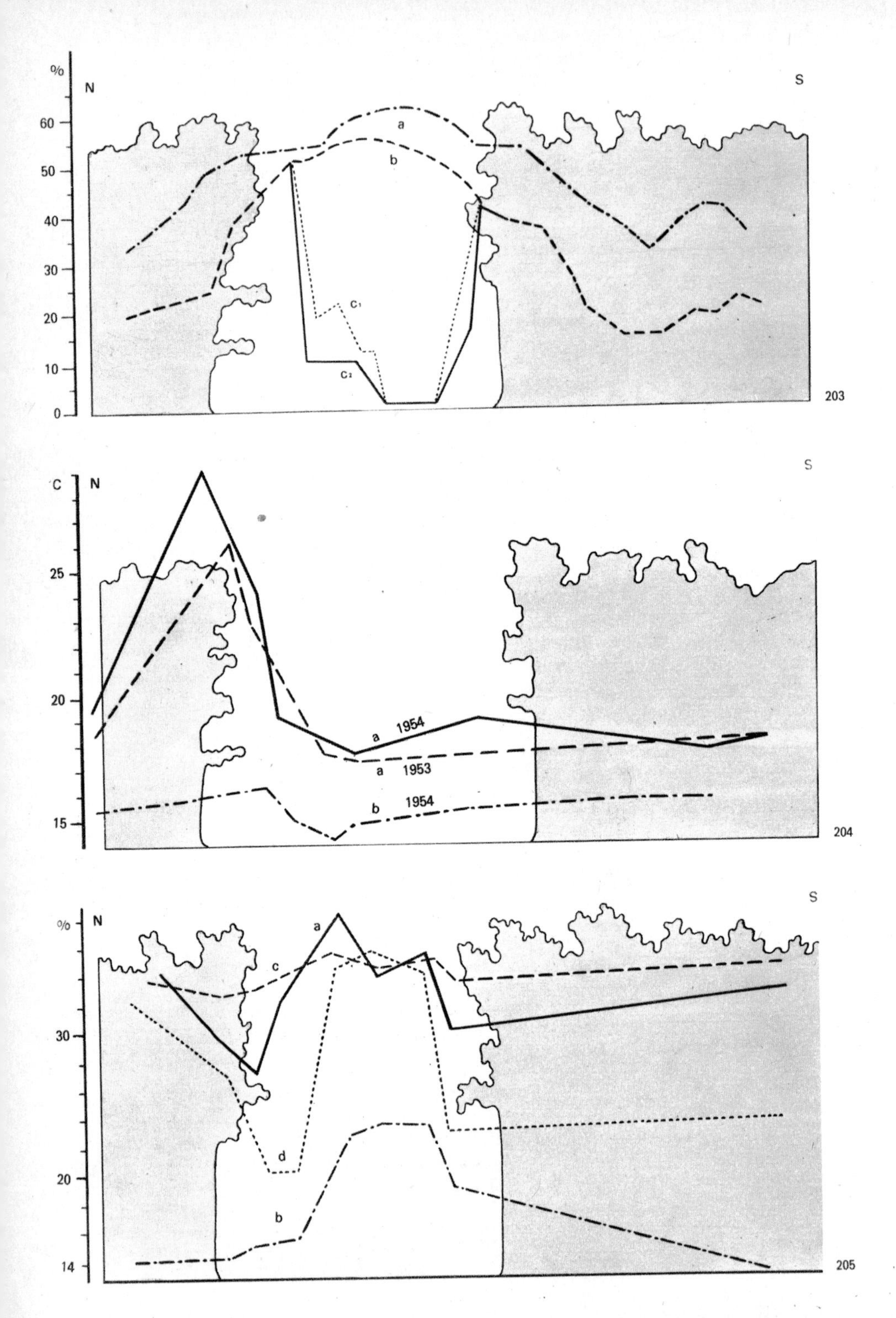
%
N
S
60
50
40
30
20
10
0
a
b
c1
c2
203
C
N
S
25
20
15
a 1954
a 1953
b 1954
204
%
N
S
a
c
30
d
20
b
14
205

206

density of the foliage, the thickness of the leaves and the form of branching. For example, a canopy of spruces and firs may reduce sunlight by 70—99 per cent, whereas under a canopy of Scots pine light is reduced by only 60—80 per cent. Also important is the age of the forest, the density of the canopy, the season and the time of day.

When a tree dies or is felled it leaves an open space which is far better illuminated than the rest of the forest (Fig. 203). This results in drastic changes. In such a clearing humus decomposes rapidly; the seeds of woody plants germinate more readily; seedlings that before had barely survived grow more rapidly; animal life in the soil as well as birds and mammals show increased activity. Many of these changes are caused not only by the increased light and the concomitant changes in temperature, but also by the changes in soil moisture and the availability of nutrients (Fig. 204 and 205). A dead or felled tree ceases to absorb food and water through its roots so more is made available to the neighbouring trees as well as to the new plant growth in the

203 Percentage of full sunlight on the north-south section of a cleared gap in a broadleaved forest; a—above the herb layer before the trees are in leaf in spring, b—above the herb layer when the trees are in full leaf in summer, c_1—under a loose herb layer, c_2—under a dense herb layer the following year.

204 Average integrated temperature on the north-south section of a cleared gap; a—on the soil surface, b—at a depth of 5 cm.

205 Soil moisture (percentage by weight of water in fresh soil) on the north-south section of a cleared gap in a broadleaved forest at various times of the year: a—16 June, b—29 June, c—27 July, d—1 September.

206 The forest microclimate always differs from that of open country and becomes 'visible' only in a special light.

207

clearing. Light varies not only in intensity and duration but also in quality, that is the proportions of red, yellow, blue, and so on that make up the spectrum. Whereas in the depth of the forest plants on the forest floor receive mostly green light filtered through the foliage of the trees, plants in clearings receive light containing equal proportions of the spectral colours.

The effect of temperature

Visible light forms part of the spectrum of electromagnetic radiation reaching the earth, and is the part vital to the process of photosynthesis. Infrared and ultraviolet wavelengths, as well as visible light, are absorbed by soil and plants, and are changed into heat which is important to the life of all forest inhabitants (Fig. 208). All important life functions such as photosynthesis, protein synthesis, respiration, growth and development take place only within a certain temperature range. Plants and animals can only survive at more extreme temperatures by means of special structures or physiological adaptations. The temperature in a forest is not constant; it changes during the course of the day and night as well as during the course of the year. The most constant temperature is found in the lower strata of tropical forests where it remains close to 25 °C, which is probably the ideal temperature for life.

The temperature at treetop level does not remain constant either in the tropics or in temperate regions. In the high mountain mist forest of the tropics every day is like summer and every night like winter— in other words, there are great differences in temperature during a 24-hour period but relatively smaller variations during the course of a year. Temperatures of about 10 °C are harmful to many tropical plants and animals, causing cessation of growth and in some instance even death.

At greater latitudes the temperature of the forest ecosystem is modified by the angle and direction of slope (Fig. 210 and 211). In North America and Europe slopes facing south-west, south and south-east receive the greatest amount of solar radiation (Fig. 209), and are covered with thermophilous forest. At latitude 50° N a southern slope with angle of 30° is best. On such a slope a forest may have the same exposure to the sun's rays (insolation) as in the flat country of Mediterranean regions. This is evidenced by the distribution of many forest species like the downy oak, the manna ash, and many herbs, beetles and lepidoptera which thrive best in comparatively hot conditions. In contrast forests on northern slopes are cooler and their composition resembles forests of the north or of high mountains.

The amount of sunlight or shade influences atmospheric moisture as well as that of the soil. Southern slopes are naturally the driest and sometimes the shortage of water is so marked that the forest gives way in places to grass species related to those of the continental steppes. Thus open spaces are formed even in the midst of a landscape where otherwise the general climate and soil are favourable for forest growth.

The effect of frost

Temperature is also affected by the movement of air at ground level which is controlled by an important physical law: cold air, being relatively denser, falls to the ground and funnels into the valleys while warm air rises. In places where this cold air is trapped it accumulates and forms a lake of cold air. Trees in such places suffer from frost, are subject to poor and unusual forms of growth and cannot form closed woodlands. In the frost basins of Europe's mountains there are frequent examples of low, conical forms of Norway spruce (Fig. 212) whose annual shoots are killed each year by the freezing atmosphere. Another feature found in such conditions is the premature fruiting of small and deformed trees (Fig. 213) which are just above the surface of the lake of icy air. Fully grown spruces with well-ripened annual shoots can, of course, survive drops in temperature far below freezing point without injury. The arolla pine is also very hardy and just as well equipped by nature as the deciduous larches (Fig. 214).

The conditions and the stage of development at which forest plants and animals are exposed to frost are important. Considerable damage is caused by late spring frosts that occur when plants have put out new foliage and flowering is at its peak. Young soft tissues with incompletely thickened walls can be easily disrupted by ice which can form in the protoplasm. Damage is also caused by ice and fro-

208

207 Mist and snow in a mountain forest.

208 Treetrunks absorb solar radiation and by reflecting heat speed up the thawing of snow.

209 Southern slopes with a gradient of about 30° host thermophilous oak forests even in a landscape otherwise covered by fir-beech forest.

209

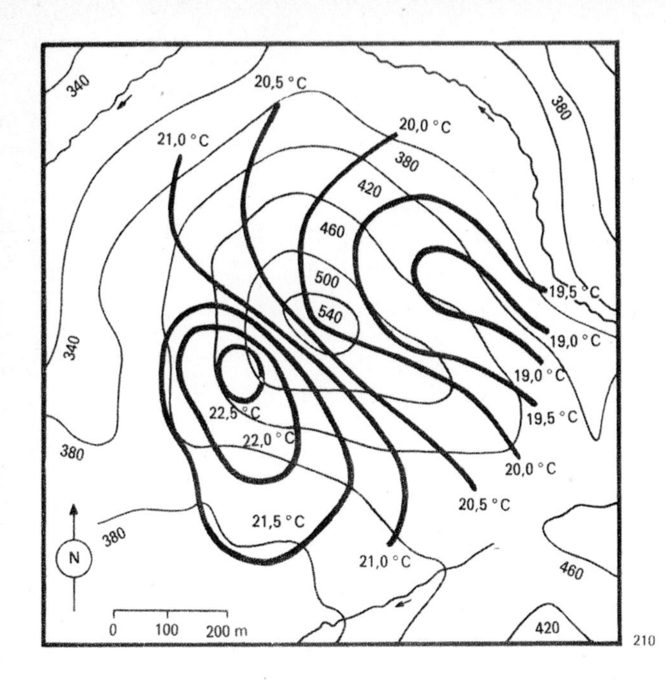

210 Distribution of maximum temperatures in the bottom layer of a beech forest on a conical hill in the course of a sunny June day; thin lines—contour lines, thick lines—isotherms connecting points having the same maximum temperatures.

zen snow on leaves and branches (Fig. 215). Sometimes a heavy, thick layer of ice forms and the tree breaks under its weight—either at some point in the crown or the main trunk.

The effect of wind

The air in and above forests is generally on the move, in the form of breezes and winds. The direction and speed of winds is determined, as a rule, by differences in pressure between warm and cool parts of the continent or between mountain ridges and valleys. To a certain degree the forest itself is responsible for the slower interchange of air between sun-warmed clearings and heavily shaded areas. A slight breeze is propitious in that it increases transpiration in the forest (and thus also the absorption of nutrients from the soil), prevents overheating of foliage exposed to the sun and disperses the carbon dioxide released into the atmosphere by the respiration of organisms in the soil (Fig. 216). Wind also plays an important role in the pollination of plants, the dispersal of spores, fruits and seeds, and also in the dispersal of many invertebrate animals—even species which are incapable of actual flight.

In regions where winds blow with great force for protracted periods the forest already has problems. In forest adjacent to prairie where there are no windbreaks in their path, winds acquire great force and exert persistent pressure from one direction on the crowns and trunks of forest stands (Fig. 217). It is easy to tell the direction of the prevailing winds by the inclined trunks as well as by the one-sided development of the crowns. On seacoasts the effects of wind on trees and forests are particularly striking. The crowns may be markedly asymmetrical and the branches bent so that they all point in the direction of the prevailing wind (Fig. 166). The zones located between 10° and 20° north and south of the equator are regions of regular cyclones (typhoons, hurricanes, tornadoes) which often destroy many square kilometres of forest leaving the landscape changed beyond recognition for years. In places where cyclones are more frequent a very dense type of forest, interlaced with lianas which offer support against wind action, has evolved.

High mountainous areas are subject to violent winds, for not only are the air currents stronger here, but their force is increased still further by the mountain topography. At the upper forest limit windblown trees with boles bare on the windward side and branches

211 Distribution of minimum temperature in the bottom layer of a beech forest on a conical hill; thin lines—contour lines, thick lines—isotherms connecting points having the same minimum temperature.

212 Dwarf spruces in a frost cauldron.

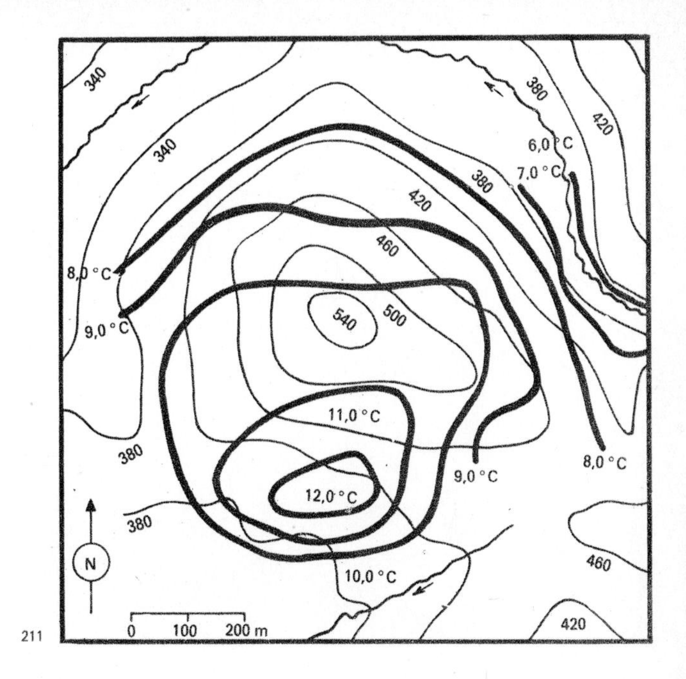

211

212

213

213 Fruiting spruce on a site affected by low temperatures, ice accretion and wind.

pointing in a leeward direction are a common sight in mountains. A forest that has grown to maturity comparatively undisturbed may suddenly be exposed to gale force winds (perhaps even from an unaccustomed direction) and within seconds it is torn up by the roots or broken like matchsticks (Fig. 220 and 221). Wind breakage usually occurs when the trunk of a well-rooted tree is buffeted by violent winds (Fig. 219) whereas uprooted trees are common in shallow or waterlogged soils where the root system is insufficiently developed (Fig. 222). Over the course of time gales affect even virgin forests. In some places the surface of the soil in such a forest becomes uneven with mounds and depressions caused by uprooted trees. The root system of these trees usually lifts a disc of soil which then falls in a heap at one side of the basin-like depression from which it was removed.

The effect of snow

Wind also enhances the effect of snow on the forest. In mountains as well as in lowlands, trees may be broken or uprooted simply by the heavy burden of wet or frozen snow (Fig. 223) and destruction may be hastened by the wind. The resistance of the forest to breakage by snow is determined by its structure, which is either the result of natural selection or planned forest management. The trees that originally grew in mountains became adapted to withstand the weight of the snow and the

The coexistence of forest and river in Alaska is no idyll.

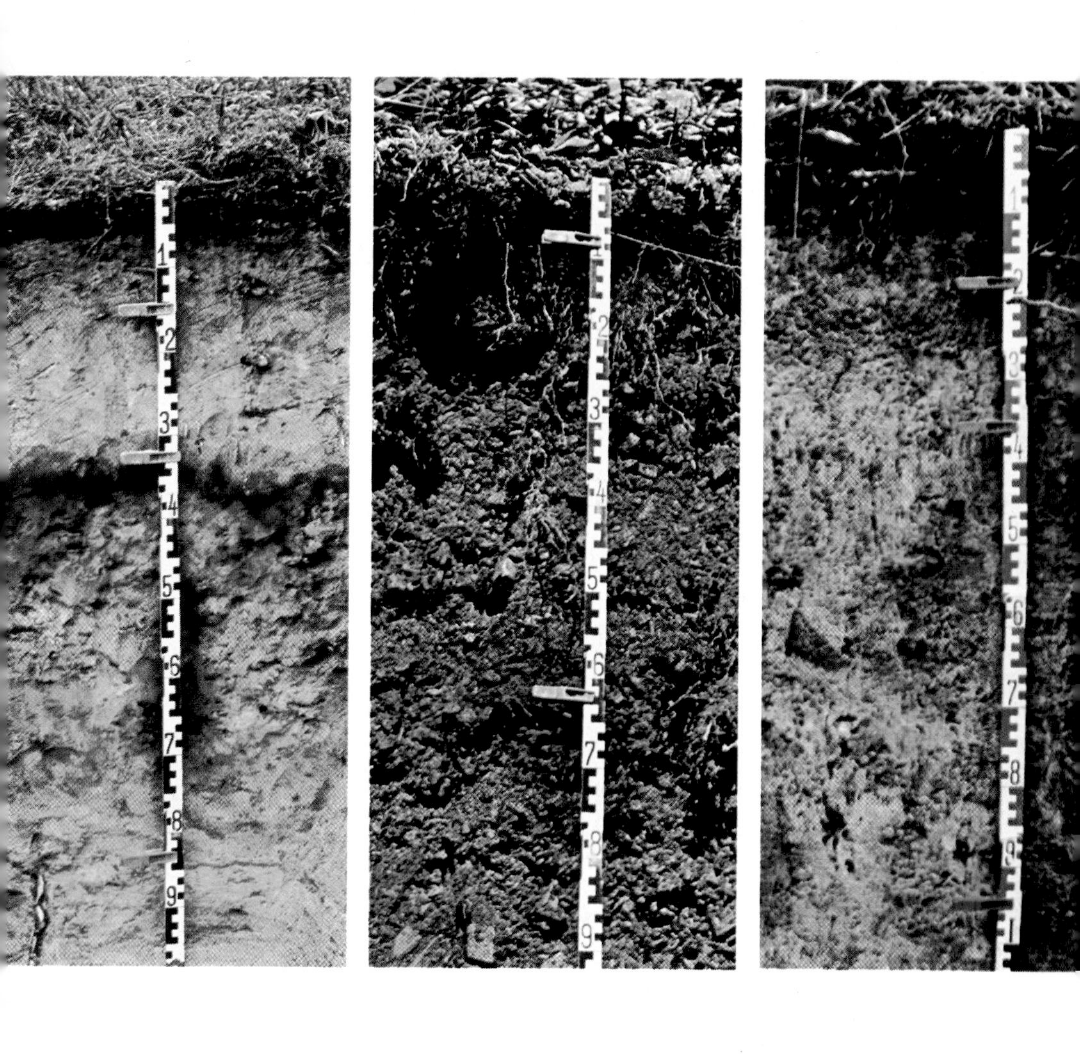

Characteristic forest soils: distinctive podzol (left), brown earth (centre) and gley (right).

214

214 Hoar frost and ice on a Norway Spruce *(Picea abies)* (left), a Common Larch *(Larix decidua)* (centre) and Scots Pine *(Pinus sylvestris)* (right). The larch is best adapted for ice accretion.

215

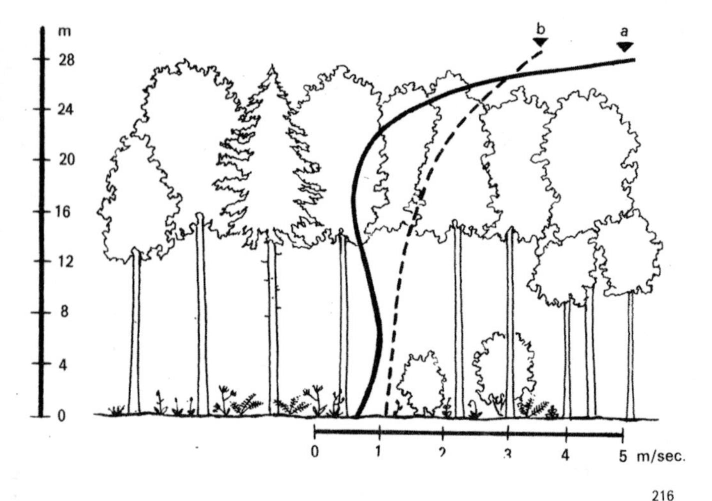

216

215 Hoar frost on the foliage of Durmast Oak *(Quercus petraea).*

216 Average speed of the wind above the forest and in the various layers: a—when in full leaf, b—when bare of leaves.

influence of winds by, for example, short, columnar boles, deep rooting systems and remarkably strong branches growing from a conical base (Fig. 225). Incorrect selection of seeds and ill-considered cultivation can result in tall, dense stands with wide-spreading branches highly vulnerable to the weight of snow and the force of gales. In central Europe the greatest damage caused by snow occurs in dense plantations of Norway spruce (Fig. 226).

Deciduous broadleaved trees, which are bare in winter and allow snow to fall to the ground instead of accumulating on the branches, are more resistant to the effects of snow (Fig. 227). The falling of snow to the ground may also influence the water balance of the forest, for more snow in the crowns means greater loss by evaporation during a thaw. Snow melts at different speeds in stands shaded by evergreen conifers and ones shaded by deciduous broadleaved trees. Here the forester has an excellent opportunity to influence the water balance of the countryside. In general, more water is conserved in forested areas than in comparable treeless areas, regardless of the type of forest. The seepage of water into the soil in spring is slower in the forest and the run-off into rivers less prone to marked extremes. These aspects of the forest are nowadays just as highly prized as the timber it yields.

As well as its weight causing mechanical damage in forested areas, snow on slopes affects growing trees by deforming the stems and curving them at the base into sabre-like shapes. As for snowslides—there the forest is completely defenceless (Fig. 229 and 230). Where snow accumulates on the steep slopes of mountains that reach above the upper forest limit, sudden devastating avalanches occur almost every year on regular avalanche tracks as well as occasionally in unexpected places. Plants growing in the paths of regularly recurring avalanches are adapted to the conditions. Fragile and brittle conifers are replaced by shrubby hardwoods or expanses of grass over which the snow slides without doing any damage. That is why avalanche tracks on wooded slopes are visible from afar. In places where an avalanche follows a course that takes it into

217

218

217 Poplars *(Populus)* bent to one side by the force of the prevailing winds.

218 Flag-form spruces on the alpine timberline.

219 A wind-broken spruce trunk after a gale.

220 The result of a strong gale in a cultivated forest.

221 Wind breakage in a spruce forest.

219

220

fully grown forest, the results are disastrous. The weight of the snow and the pressure of air round it breaks or uproots all the mature trees and carries them to the bottom of the valley.

The effect of the soil

The characteristics of a forest are influenced also by the nature of the soil — its depth, porosity, fertility, acidity and moisture and also the soil organisms it contains. The quality of soil is determined by the mother rock in the substratum, and the effects of the forest ecosystem in the area. There are many different kinds of rock on the earth, and their various physical and chemical characteristics may be greatly altered in the case of very old soils by the long-term effects of the climate and the living part of the ecosystem. Some rocks, however, have a marked influence on the process of soil formation. Rocks composed of carbonate compounds (limestone, dolomite) differ greatly, as a rule, from siliceous rocks (granite, gneiss, phyllite, sandstone) in the quantity of available nutrients and water made available, as well as in the type of humus and general fertility. Equally marked is the influence of certain species of forest trees, which affect the soil by their root systems (Fig. 228) as well as the quantity and quality of their litter (leaves, twigs). Its beneficial effect on soil has earned the European beech the nickname 'mother of the forest'. North American hardwoods that could be similarly named include not only the related American Beech *(Fagus grandifolia)* but also the widespread sugar maple. Most conifers, on the other hand, do not have a beneficial effect for they promote the formation of raw humus, and give rise to leaching, and increased acidity. Exceptions, however, prove the rule, as in the case of the silver fir, which, particularly in combination with beech, produces fertile soil (Fig. 232).

221

The chief cause of difficulties in the cycling of nutrients in the forest ecosystem is the formation of raw humus, called mor. It is formed during the process of decomposition when leaf proteins are precipitated by the action of phenols released from the needles and leaves of certain trees. Protein complexes themselves resist decay and furthermore protect cellulose,

222

222 Uprooted tree with spreading (plate-like) root system.

223 Snow and ice accretion leave permanent traces on the shape and growth of trees.

224 Wind is often the last factor in the destruction of an old tree.

224

the principal component of litter, from decay as well. Products of raw humus influence the transfer of nutrients to the soil depths, increase the acidity of the soil, and generally reduce the activity of soil bacteria and fungi. In time raw humus is covered by a continuous carpet of mosses and lichens (Fig. 231), which prevents the germination and growth of tree seedlings. This may mark the last stage in the degradation of the forest.

The forest is a very adaptable ecosystem and grows on a wide range of soils to which the soil scientists have given specific names. Brownearth is deep fertile soil composed of layers coloured different shades of brown merging imperceptibly with one another. Podzol is acidic soil composed of distinct layers between which is a middle layer of fine earth coloured ash grey. Rendzina is humus-rich soil forming a homogeneous layer on calcareous rock. Ranker is humus-rich soil forming a ho-

223

225 Clearly visible on the dead trunk of a Norway Spruce *(Picea abies)* are the conical bases of branches — an adaptation to the weight of snow and the buffeting of winds.

mogenous layer on rock poor in calcium. Gley is moist and temporarily waterlogged soil with a blue-green layer of subsoil which is deficient in oxygen. The effect of these soils on the plant cover depends in great measure on the size of the mineral particles, in other words, on the amount of clay, silt, sand, gravel and larger stones they contain. An extreme type is scree soil which is formed in mountains mostly as a result of frost weathering and which is host to special types of scree forests. Between the boulders of the scree there are always areas of fine earth containing minerals and humus so that the various forests can contain even demanding species of trees and soil organisms. Quite different is the rocky soil in a temporary or former riverbed (Fig. 233), which contains no fine earth or humus whatsoever and where no proper forest can ever gain a foothold.

Comparatively fertile soils have formed in regions that were once covered by the ice-sheet or a valley glacier. Bare and partly dispersed moraines (Fig. 234) are composed of large boulders and fine earth which is a sufficient source of nutrients and is also the basis for vigorous forests. Such soils are found in the north of Europe (including the northern parts of both East and West Germany) and northern North America (including a large part of New England). Deep sandy soils (Fig. 236) allow roots to penetrate to great depths but because of insufficient nutrients and poor water retention they are generally covered only by undemanding stands of pine. Tree roots find it difficult to penetrate crevices in granite (Fig. 235) but the environment at the foot of a cliff or rocky slope may be congenial for a forest because the soil is continually enriched by nutrients from weathering minerals and kept moist by the run-off of water.

The forest, on the other hand, is an important factor in the battle against erosion on slopes, on the shores of lakes (Fig. 237) and on the coasts (Fig. 238). Its protective influence is desirable wherever strong water currents and fierce winds hasten soil erosion and threaten to denude the landscape. This function of the forest has regrettably been put to the test by man on many occasions (Fig. 239).

Forest grows even on soils which display very unusual chemical compositions. These include, for instance, serpentine soils that have an extremely high content of magnesium and heavy metals (chrome and nickel) and an extremely low content of important nutrients

226

227

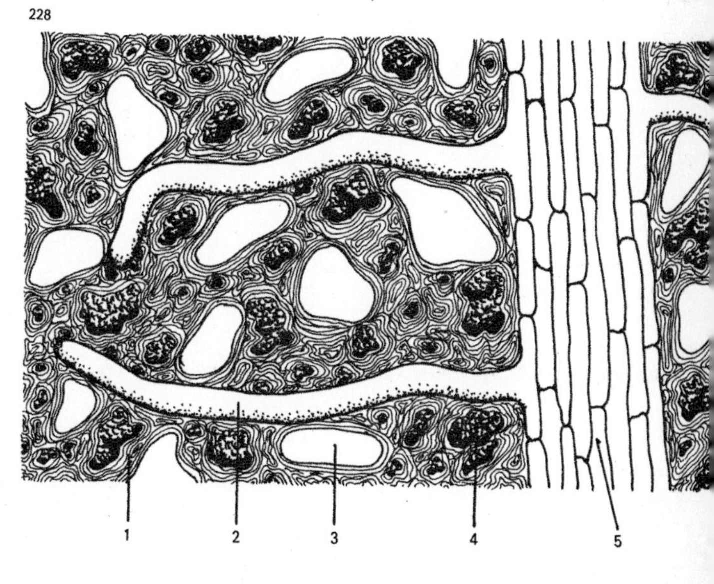

228

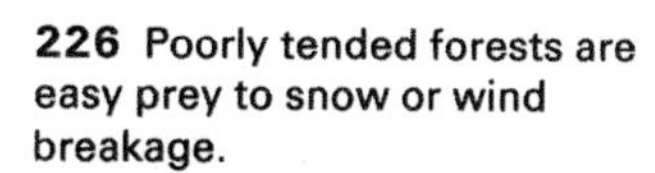

226 Poorly tended forests are easy prey to snow or wind breakage.

227 Deciduous broadleaved trees allow snow to fall to the ground.

228 Structure of the soil in the vicinity of an actively absorbing tree root: 5—root, 2—root hairs, 4—mineral soil particles, 1—soil water, 3—air-filled pore.

229

230

229 A periodic avalanche path permanently covered with shrubs cuts into the forest complex.

230 A tangle of trees in the runout zone of an avalanche.

231 A carpet of moss *(Leucobryum glaucum)* spreads over the raw humus under a spruce monoculture.

232 The soil in mixed forests of European Silver Fir *(Abies alba)* and Common Beech *(Fagus sylvatica)* is good soil (brown earth) because of the beneficial influence of the decomposing litter.

(calcium, phosphorus, and nitrogen). In Europe, forests on serpentine soils consist chiefly of Scots pine, which repeatedly surprises with its minimal requirements and resistance to toxic influences (Fig. 240).

A distinctive kind of environment is that found in peat bogs (Fig. 242) where the soil is composed almost solely of organic remains of plants such as bog mosses. The accumulation of peat is caused by the absence of bacteria and fungi, which normally break down plant material. This is due to a deficiency of the basic nutrients and sometimes even of oxygen which such bacteria require. Only specially adapted undershrubs and shrubs of the heath family can grow among actively growing peat mosses and peat sedges (Cyperaceae). However, at the edge of the peat bog (Fig. 241), where the water table is near the surface and plant roots can reach the mineral substrate, pioneer conifers (pines, spruces) as well as broadleaved trees like the common birch are found.

Until comparatively recently it was assumed that peat forests occur only in the temperate zone but it is now known that there are peat deposits with their characteristic forests even in the humid tropics of South-east Asia and South America. In Borneo the dominant tree of such forests is *Shorea albida* of the Dipterocarpaceae family, which grows to a height of fifty metres. In the Amazon basin there are extensive peat forests in the vicinity of the so-called black rivers, (e. g. Rio Negro).

231

It is not unusual for trees to grow on artificially produced soil as, for example on quarry heaps containing rocks brought up from the depths, which at first lack the fundamental characteristics of fertile soil. In Europe the tree most often found growing in such adverse soil conditions is the birch (Fig. 244). Forest spreads even over the ruins of old castles (Fig. 243). After ten or twenty years the partly crumbled walls and rubble are concealed by a continuous cover of European Elder *(Sambucus nigra)*, willow, birch and pine.

232

233

234

233 In the boulder-strewn bed of a mountain stream, growth of trees is prevented by floods and lack of fine earth.

234 Moraines are a mixture of boulders and fine earth; their forests suffer from water erosion.

235 Weathered granite rock with pioneer spruces.

236 The exposed roots of a Scots Pine *(Pinus sylvestris)* on the wall of a sandstone quarry bear witness to the extensive root system of trees.

235

236

The effect of fire

Forests the world over are continually threatened by fire. Forest fires are caused not only by man, as described in Chapter 2, but are also a natural phenomenon that has occurred since time immemorial, long before the forests became inhabited by man. Burned remnants of wood may be found even in the depths of old sediments and many structural and functional adaptations of trees in savannas and dry forests cannot be explained otherwise than as the result of the selective effects of fire over hundreds of thousands, or millions of years.

The principal cause of forest fires in virgin land uninhabited by man is lightning (Fig. 245). It was formerly assumed that forest fires were caused by lightning only on rare occasions because thunderstorms are usually accompanied by heavy downpours which quench the fires. Recent data, however, indicate that storms without rain are quite common and that a forest that has been set on fire may continue smouldering during a downpour, the fire breaking out anew when the ground has dried. In Nebraska thirty fires caused by lightning broke out in a single day when climatic conditions were favourable; five of these attained such vast proportions that they caused a million dollars' worth of damage. Near active volcanoes forest fires are caused by hot lava and ashes that are deposited on the slopes after a sudden eruption. On rare occasions forest and savanna fires have been known to be caused by rocks falling

237

down steep mountainsides for these, if they are sufficiently hard, produce sparks that can ignite dry grass or litter. To date there is no evidence to support the possibility of spontaneous combustion, which could occur in places where a thick layer of damp wood or other plant material has accumulated, as a cause of forest fires although instances of spontaneous ignition in piles of wood chips in the woodworking industry are not unknown.

In inhabited regions the commonest cause of forest fires nowadays is man and his technology. For example, in 1959 fire destroyed 7,636 hectares of forest in the German Federal Republic, and this was caused in ninety per cent of the cases by man and machines. As the number of hikers and holiday-makers increases in North America and Europe so does the number of fires caused by burning cigarette ends, sparks from camp fires, and matches in the hands of children. In countries where steam engines are used on the railways they are a constant source of danger, particu-

238

237 The tangle of forest tree roots prevents erosion of the shore by the surf.

238 The Red Mangrove *(Rhizophora mangle)* holds the soil on Cuba's eastern coast by means of its stilt roots.

239 Washed-down soil and ravines below a mountain slope, deforested without due consideration.

240 Scots Pine *(Pinus sylvestris)* on serpentine soil.

239

larly in spring when the old grass on the forest floor has become dry.

Any forest in any part of the world may be the victim of fire. Accumulated organic matter burns even under conditions that were initially moist. The progressing line of fire gradually dries out even the wettest matter and heats it to ignition point. The wood and bark of conifers which contain resin burn best. In tropical rain forests trees that have been girdled several weeks previously and whose weaker branches and leaves have dried slightly, burn well.

Noted for their dimensions and destructiveness are the huge fires of the coniferous taigas of the northern hemisphere. The forest fire that broke out in central Siberia in 1915 encompassed an area equal to about a third of Europe. Despite costly protective measures even Canada has its share of fires every year, which devastate thousands and even tens of thousands of hectares. Alaska is another state hard hit by fires. The arid climate of this region coupled with the lengthy days of summer provide ideal conditions for the outbreak of fires. In Scandinavia spring and summer fires are so common that foresters and ecologists consider them the chief factor influencing the distribution of spruce and pine forests. In Lapland there are many instances of pine forests

whose ages correspond to large forest fires which are known to have occurred 400, 300, 150 and 70 years ago. In the cultivated coniferous forests of central Europe and North America fires are a constant threat. For example, in the summer of 1969 there were 2,829 fires in Poland which destroyed 5,300 hectares of forest!

Vast areas are destroyed by fire in subtropical and monsoon forests where part of the year is without rainfall. Fires in the savanna woodlands of Africa, South America and Australia are so frequent that it is assumed the same area is affected about once in two years.

241

241 Trees growing in soft peat are not firmly anchored and thus tend to lean to one side.

242 The trees best suited for peat bogs are pines and birches.

242

The undemanding Scots Pine *(Pinus sylvestris)* grows in the harsh climate and poor soil of an island in the North Sea.

Only the specialized pine *Pinus rotundata* and Scots Pine *(P. sylvestris)* grow on the edge of an active peat bog.

The huge fire that broke out in the state of Paraná in Brazil in 1963 destroyed five million hectares of forest in one sweep. Proof that even the moist tropics are not safe from the threat of forest fires are the lightning gaps in the swamp forests of Borneo. These are small spaces left by burned forest in the heart of uninhabited territory which have been revealed by aerial photographs. In regions with a Mediterranean climate the critical period for forests is the dry summer season. Responsible for the devastation of the Balkans, in addition to excessive logging and ill considered pasturage, are destructive forest fires which are aggravated by the mountain topography that acts like a chimney with a good draught.

Countries where the danger of forest fires is greatest, like the Soviet Union, Canada and the United States, have established a branch of forestry that is concerned with the prevention of forest fires. They have organized a vast system of observation posts and have fire brigades with parachutists, modern communication systems, computers, aircraft, sand-throwers, chemical fire-extinguishers, and fire-extinguishing bombs at their disposal. Research insitutes in Canada study the processes of burning in forests and examine the possibilities of preventive control by means of intentional burning of combustible brushwood and litter.

The forest is marked by constant change.

243

243 Sooner or later even old castle ruins are overgrown with forest trees and shrubs.

244 A heap of waste rock from a surface mine is slowly stabilized by Silver Birch *(Betula pendula)*.

244

245

245 A treetrunk scarred by lightning.

246 Summer in a mountain spruce forest.

247 Winter in a mountain spruce forest.

During the course of the year the temperature changes, as does the moisture, snow cover, density of foliage, colour of the leaves and fruits, and activity of the forest wildlife. In temperate regions there are marked contrasts between the summer and winter appearance of the forest (Fig. 246 and 247)—two entirely different stage sets with actors in different costumes going through different motions and producing different sounds and scents. Even in tropical jungles there are seasonal changes responding to less perceptible environmental factors such as changes in the length of day. The fall of an old tree in a forest in any part of the world is the beginning of marked changes in the surface climate and soil forming processes.

As the environment changes so does the forest and as the forest changes so does the environment.

246

247

248

248 Sunlight is the main source of energy for all creatures in the forest ecosystem.

THE WEB OF LIFE

The uninformed person may see only trees in the forest, and many a forester looks upon his territory as merely a collection of trees and a source of timber. But the forest is much more than that. It is a complex organized natural or ecological system—an ecosystem. In the forest ecosystem all the living plants and animals are inseparably bound to one another as well as to the environment by invisible paths of energy flow and material cycling systems. Just as important as the trees in the life of the forest is the diversity of organisms with their differing life requirements. Forest organisms may be simply classified as producers, consumers and decomposers.

Sources of energy
The sun

Trees and all other green plants are producers of organic matter. With the aid of light energy from the sun they transform simple chemical substances in the environment (chiefly water and carbon dioxide) into organic substances (sugars, fats and proteins) in their tissues. Consumers comprise herbivores, which are animals that feed on green plants, and carnivores, which feed on other animals. Absolutely essential to life in the forest are the decomposers that use dead plant and animal material as food, breaking down the organic substances and releasing simple raw materials into the environment. Decomposers include animals such as worms and insect larvae which feed on dead and decomposing organisms, and bacteria and fungi which further aid the breakdown of organic material into simple substances. It is clear that without decomposers life in the forest would soon come to a dead end. Viewed objectively, then, small mites and microscopic bacteria that break up cellulose are just as important in the forest ecosystem as a large tree (Fig. 249).

The common fate of forest organisms depends on the supply of solar energy required by the green plants for photosynthesis.

Photosynthesis

Starting substances	6	molecules of carbon dioxide
	12	molecules of water
	2862	units of energy (joules)
Complex reactions		in the presence of light and chlorophyll
Final products	1	molecule of glucose
	6	molecules of oxygen
	6	molecules of water

Light in the forest is available only in limited quantity and for varying lengths of time according to the period of daylight and the season of the year (Fig. 248). In a closed forest with a dense canopy the intensity of life within the forest is always restricted by lack of light. During their lengthy evolution forest plants have become adapted so as to make the best use of the space and light available. It is this which gives the variety of adaptations in the body structure and functions which are characteristic of forest plants. Natural forests are thus generally composed of many species of woody plants and include individuals of varying ages and life forms. This permits their leaves to utilize every glimmer of light in the space above the surface of the soil (Fig. 251). The resultant layered arrangement of leaves, branches and trunks, or stratification, of forests leads to the economic utilization of solar energy and puts forests among the most productive areas on the earth (Fig. 250). The coniferous taiga can produce 400 to 2,000 grams (dry weight) of organic matter per square metre a year, a deciduous forest 600 to 3,000 grams per square metre and a tropical rainforest 1,000 to 5,000 grams per square metre. In virgin forests that are not logged this potential production is, of course, not utilized, and the organic material formed is recycled within the ecosystem (Fig. 254).

Other aspects of the environment that are also important for optimum utilization of solar energy include moisture and the fertility of the soil. The diversity of forest plants, both as re-

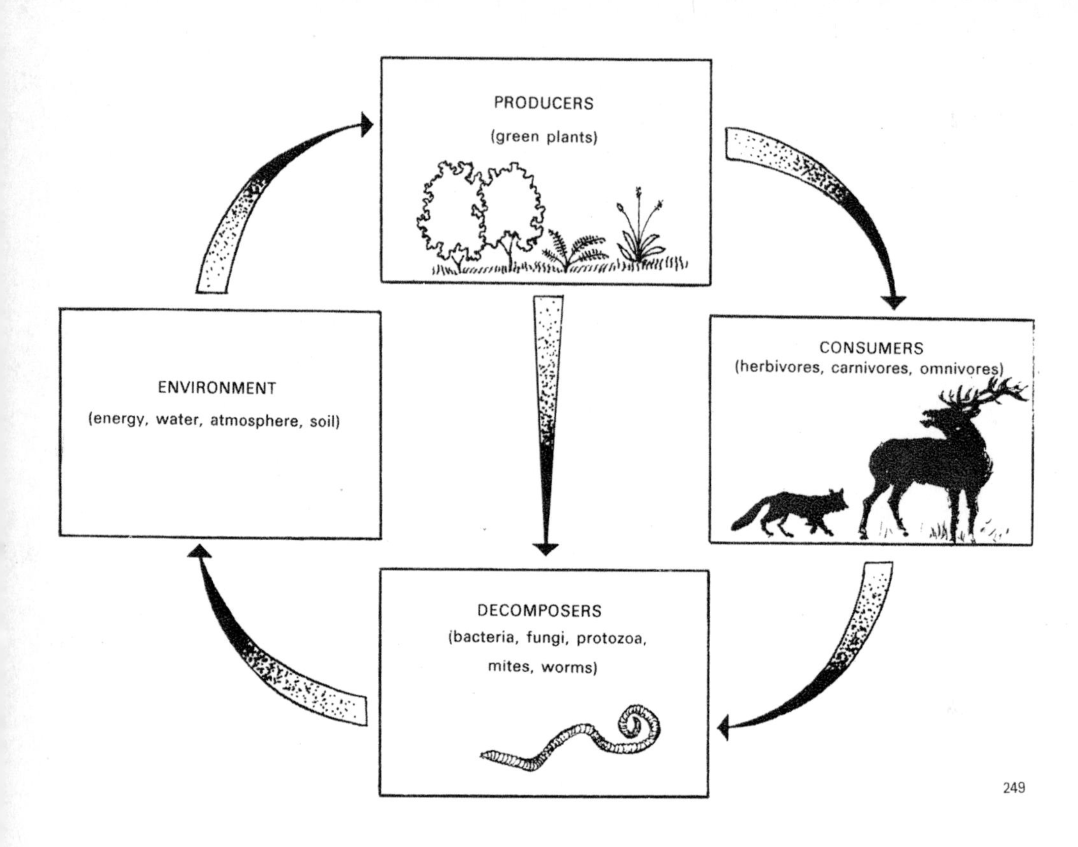

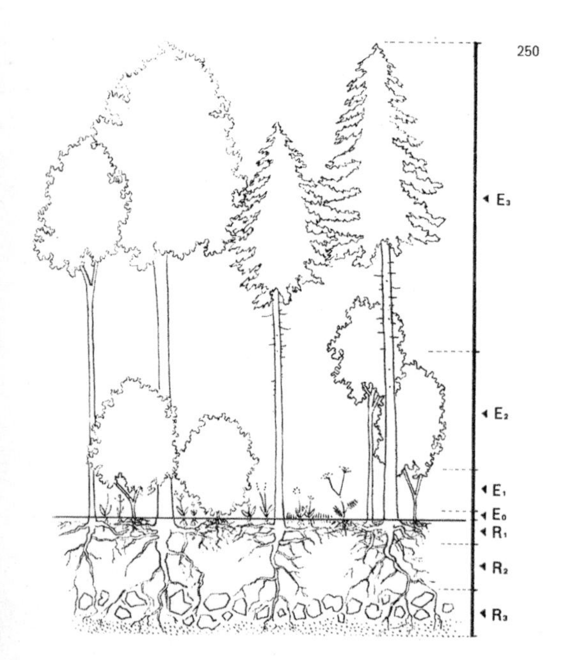

249 Diagram showing the food and energy relationship in the forest ecosystem.

250 Stratification of the forest above and below ground: E_0 — moss layer, E_1 — herb layer, E_2 — shrub layer, E_3 — tree layer, R_1 — upper root layer, R_2 — middle root layer, R_3 — lower root layer.

251 A mixed beech-fir virgin forest makes efficient use of sunlight.

252 A young forest begins life in a small clearing.

gards age and species, contributes to better utilization of all natural sources. Various generations of trees occupy various layers in the space above as well as under ground and interlace to form a complex mosaic of full-grown trees, young trees and gaps with germinating seedlings and new growth (Fig. 252). This is also characteristic of the flood plain forests of the fertile large-river valleys in the temperate zone. The perpetually high water-table and fertile alluvial deposits give rise to a vigorous community of both sun-loving and shade-loving trees, lianas, shrubs and herbs (Fig. 255) and equally varied animal life.

The space above the ground is superbly utilized in tropical rainforests (Fig. 253) where giant trees, that rise above the average level of the canopy, live side by side with trees of moderate height that form the dense middle tier, and dwarf trees that live permanently in the shaded depths below. Tropical forests also include a great number of lianas that climb up the trunks of robust trees so that their light-demanding leaves and flowers can make the best use of every gap in the canopy where the light is stronger. The small lianas of the family Araceae are common in all tropical forests;

251

252

253

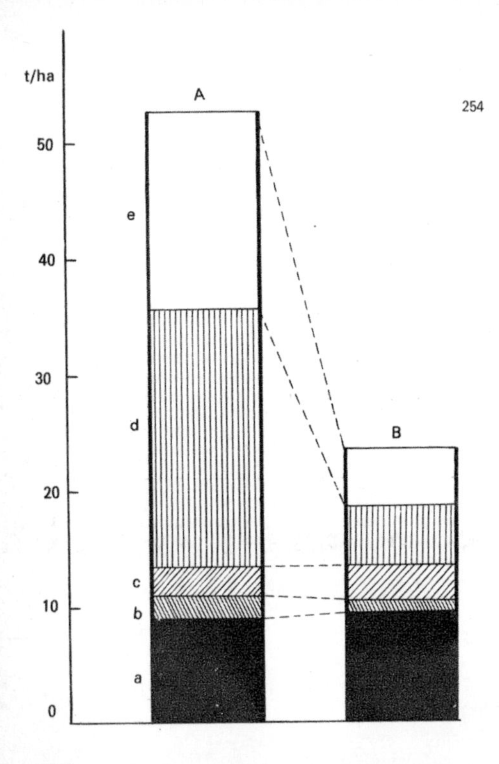

253 The layered structure of a tropical rainforest in west Africa.

254 Comparison of seasonal changes in the biomass of (A) tropical and (B) a temperate broadleaved forest measured in tonnes per hectare: a—net annual production of roots, trunks and branches, b—dead roots and branches, c—dead leaves, d—loss caused by respiration of roots, trunks and branches, e—loss caused by respiration of leaves.

255 Forest space full of live branches and green foliage.

256 The furrowed stem of the liana *Neuropeltis prevosteoides* climbs to the topmost tree of a virgin rainforest; West Africa.

less common are the woody lianas, for example of the Convolvulaceae family, with stems up to thirty centimetres thick and more than a hundred metres long (Fig. 256). The tropical forest also includes herbaceous plants of various heights, some being as much as five metres tall.

Different in its structure and adaptation to light is the coniferous taiga of northern Eurasia and North America. In primeval forests, spruces, firs and hemlocks form a closed canopy that allows little light to penetrate. All other life below is slowed down, and even the lower branches of the dominant trees die. This results in a seemingly empty space filled only with the rigid columns of tree trunks. In shaded taigas only very hardy and undemanding lichens, mosses and herbaceous plants are found on the forest floor. Coniferous forests composed of pine trees allow more light to penetrate and permit the development of denser undergrowth. In rather dry climates and on poor soils light-loving pines can grow successfully even alongside spruces, and form a broken canopy (Fig. 257).

In the forest ecosystem light is neither constant nor always of the same intensity. It is limited to the period of daylight, which changes in length during the course of the year. Only in the equatorial rain forests is the supply regular, with twelve hours of light and twelve of darkness daily (Fig. 258). The northern taigas have as much as twenty hours of daylight in summer and only a few in winter. In periods of inclement weather (drought or cold) many broadleaved trees and some conifers shed their leaves. This changes the light conditions of the entire forest ecosystem (Fig. 259 and 261). The leafless trees allow sunlight to penetrate to the ground and with the arrival of warm spring weather growth is recommenced and the physiological functions of all life on the forest floor are reactivated. The flowering plants that are so characteristic of Europe's and America's forests in spring are but one sign of the many activities taking place within the forest at this season.

The importance of light to forest life is particularly noticeable at the edges of the forest. The trees here are more branched and have more massive crowns. A mantle is soon formed of herbaceous plants, lianas and trees which take advantage of the light coming from the side (Fig. 260). *Cecropia* in the tropical forests of America and *Musanga cecropioides* (Fig. 262) in the forests of Africa are frequent

255

256

257

257 Canopy of a coniferous forest with Scots Pine *(Pinus sylvestris)* and Norway Spruce *(Picea abies)* viewed from the ground.

258 There are only 12 hours of light on the ground stratum of a tropical virgin forest and even less under dense crowns.

259 The interior of a deciduous forest with Common Beech *(Fagus sylvatica)* and Silver Birch *(Betula pendula)* in spring.

components of such a mantle. When light falls on the soil there is rapid germination of seeds which may have lain dormant for many years in the shaded humus of a fully grown forest. In full sunlight *Musanga cecropioides* can grow to a height of fifteen metres in five years, which is quite a commendable feat in the plant kingdom. Pioneer trees in the temperate regions of Eurasia and North America also grow rapidly. These include various species of birch, poplar, willow and alder. The need to make

258

259

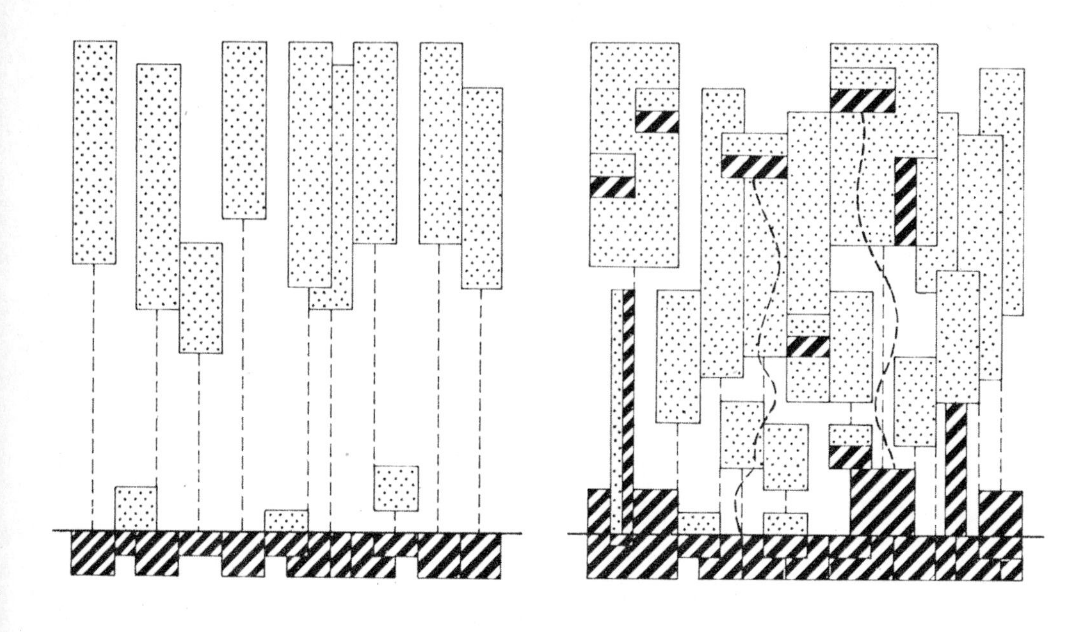

260 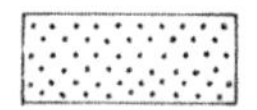 leaves  roots

261

maximum use of all available light is evident both in the arrangement of leaves in space (leaf mosaic) and the general shape of the crowns. Emergent tropical trees have spherical (Fig. 263) or broad umbrella-shaped crowns. The oaks of the temperate regions make broadly-spreading crowns in full sunlight, this being more easily observed in winter (Fig. 264). Groups of boles that have grown from a stump come in time to look like an open rosette that spreads its branches and leaves so as to receive the greatest amount of light (Fig. 265).

The flat leaf blades of broadleaved trees are arranged so that their upper surfaces utilize direct overhead light and the undersides make use of the energy from weaker, dispersed and reflected light. The narrow needles of conifers do not always assume an organized position in space but capture light from all sides (Fig. 266). The quantity of light also affects the multiplication of forest plants and animals. Flowers are generally produced on branches exposed to plenty of light, which also assures

260 Arrangement of green leaves and active roots in (left) a European broadleaved forest and (right) a tropical rainforest.

261 Herbaceous plants in spring in a flood plain forest before the trees come into leaf.

262

263

262 Sunlight is fully utilized by pioneer trees at the margin of the tropical virgin forest; *Musanga cecropioides* in west Africa.

263 Spherical crowns of tropical rainforest giants; at left—*Celtis mildbraedii,* at right—*Triplochiton scleroxylon;* Ghana.

264 Broadly-spreading crown of an English Oak *(Quercus robur)* in a flood plain forest.

265 Trunks of coppiced lime trees grow in fan-like form so that the branches and leaves receive as much light as possible.

264

265

266

267

266 **The needles of conifers are arranged in many different positions and are ready to start photosynthesis as soon as the frost is over.**

267 Most forest trees produce flowers and ripe fruits only on branches exposed to full sunlight; Common Beech *(Fagus sylvatica).*

proper maturation of the fruits (Fig. 267). Seeds and fruits contain stores of food (and thus of energy) necessary for the growth of new seedlings (Fig. 268).

In parts of the forest where there is ample light small plants may have difficulty in competing with other plants of the ground layer, owing to the slowness of their growth. Such plants may adapt by carrying out photosynthesis at a generally lower rate, while others, called epiphytes, grow upon other plants of the ecosystem. In forests throughout the world green algae, lichens, mosses, liverworts and small ferns (Fig. 270) grow on the trunks and branches of trees, and in tropical forests algae, lichens and mosses commonly cover even the foliage of trees. More robust herbaceous plants such as orchids, bromeliads and ferns may occur in the forks of branches of tropical trees (Fig. 269). Epiphytes are often succulent forms that tolerate the heat of the tropical sun as well as marked daily fluctuations in temperature and atmospheric moisture. When the host tree sheds its leaves they

Mixed forest with Norway Spruce *(Picea abies),* Silver Fir *(Abies alba)* and Common Beech *(Fagus sylvatica)* in the Bohemian Forest, Czechoslovakia, is a balanced and very productive forest ecosystem.

The Red Deer *(Cervus elaphus)* is a herbivore and as such is an important link in the forest ecosystem of Eurasia and North America.

268 The acorns of the English Oak *(Quercus robur)* contain stored energy for the germinating seedlings.

268

269

269 The epiphytic orchid *Aerangis kotschyana* is noted for its remarkably long spurs.

270

271

272

270 Whole gardens of lianas, mosses and ferns grow on the trunks of tropical trees; Atewa Range, Ghana.

271 Epiphytic orchids and ferns high up on the crown of the deciduous tree *Parinari excelsa;* when the tree is bare of foliage the epiphytes receive light from all sides.

272 The leaves of the dwarf tree *Allexis cauliflora* fan out on all sides in the dark interior of the rain forest.

273

273 Epiphytic plants utilize the light in the upper layers of the forest.

274 The mosaic of leaves of *Asarum europaeum* makes the maximum use of the light from above.

274

275 Diagram to show the energy flow in the forest ecosystem: E—energy for photosynthesis; R—heat loss to outer space; P—producers (trees, herbaceous plants); C_1—consumers-herbivores; C_2—consumers-carnivores (predators—beasts, birds, insects); C_3—consumers-higher level carnivores (predators and parasites feeding on other carnivores); D—decomposers (bacteria, fungi and other organisms participating in the decomposition of organic matter and breaking it down into mineral substances).

276 Most forest ferns tolerate shade but grow much better if they receive light from the side.

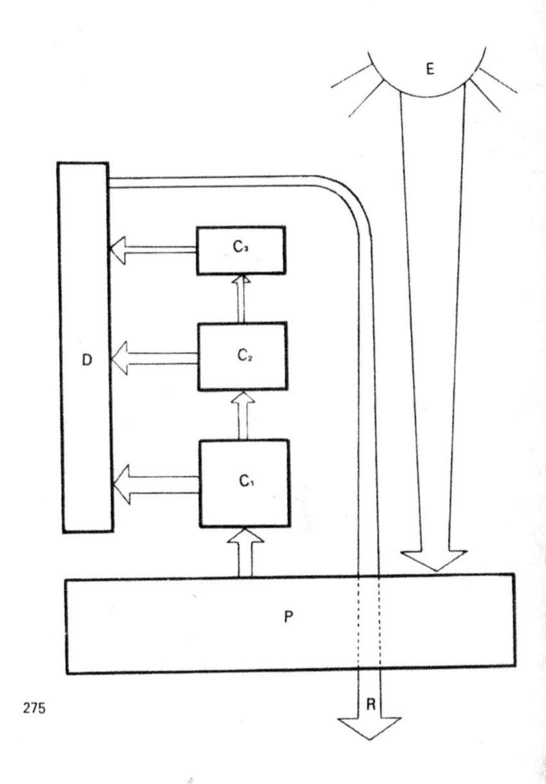

275

276

277

278

277 Sugars in the fruit of the Raspberry *(Rubus idaeus)* are an important source of energy for some forest animals.

278 The fruits of the Wild Gooseberry *(Ribes grossularia)* are a favourite food of insects, birds and mammals.

279

are exposed to a flood of sunlight from all sides (Fig. 271).

Dwarf trees, herbaceous plants and ferns that grow in the gloom of the lower tier (Fig. 272, 273, 274 and 276) have developed various structural and physiological adaptations during the course of evolution which enable them to flourish in light that is often less than one per cent of full sunlight.

The solar energy captured by photosynthesis in the green leaves of trees is mostly stored in the wood (Fig. 275 and 279), though it is also stored in the twigs, roots and bark. Also very rich in energy are the sweet fruits of the woody and herbaceous plants of the forest (Fig. 277 and 278). The further fate of this captive energy in the ecosystem depends on the activity of the consumers and decomposers. Of course at each successive stage some of the energy is lost in the form of heat for it is utilized by the organism in the process of respiration.

The soil

The diverse life of the forest depends also on the sources of raw materials, including trace elements taken in solution from the soil by trees, lianas and herbaceous plants (Fig. 280, 281 and 282). The forest ecosystem is also efficiently organized in the soil, the driving force behind this being the competition of roots for food and water. These are not equally distributed in the soil which is why roots are arranged in layers. The upper root layer is located in the top few centimetres of soil, which contains the greatest amount of available nu-

280

279 Most of the energy captured by the forest ecosystem is stored in the wood of treetrunks.

280 As well as sunlight the forest ecosystem requires water and food; these are obtained from the soil.

281

trients (Fig. 283 and 285). As much as eighty per cent of all roots, which branch to form physiologically active organs here, are crowded into the top twenty centimetres of soil. A smaller number of roots forms a middle layer, where nutrients either leached from decomposed humus or carried down by soil organisms are available. The bottom layer is formed by tap roots and the anchoring roots of large trees. If these roots reach to the water table they make an important contribution to the tree's supply of water. To reach food and water roots often extend far from the trunk and grow over and around obstacles in their path (Fig. 284).

If the ecological stability of the forest ecosystem is to be maintained, it is essential that most of the nutrients from dead plants and animals are returned to the soil. This nutrient cycle is particularly important in tropical forests where the laterite soils themselves are very poor in nutrients. Most nutrients in the tropical rainforest are stored in the living plant and animal material (the biomass) and only

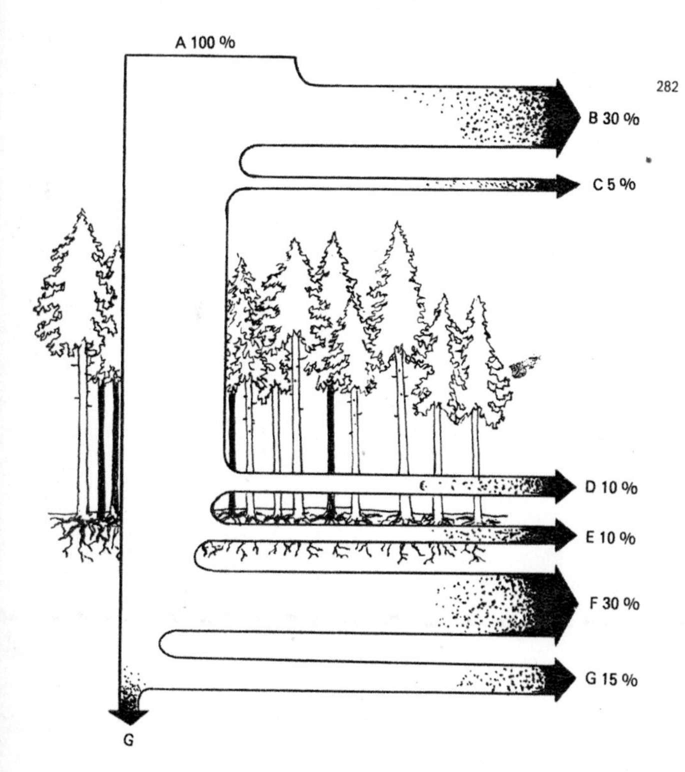

282

281 A well-branched root system provides trees with a continuous supply of nutrient solutions; roots of the Scots Pine *(Pinus sylvestris)* pulled up from the soil.

282 Water balance in the taiga: A—precipitation, B—water intercepted and evaporated in the treetops, C—surface run-off, D—consumption by the herb layer, E—sidewise percolation of water through the soil, F—consumption by the shrub and tree layer, G—run-off of underground water.

283 Calcium cycle in a forest; the numbers denote kg/hectare per year.

284 Tree roots grow across boulders and other obstacles to obtain water and nutrients; Norway Spruce *(Picea abies)*.

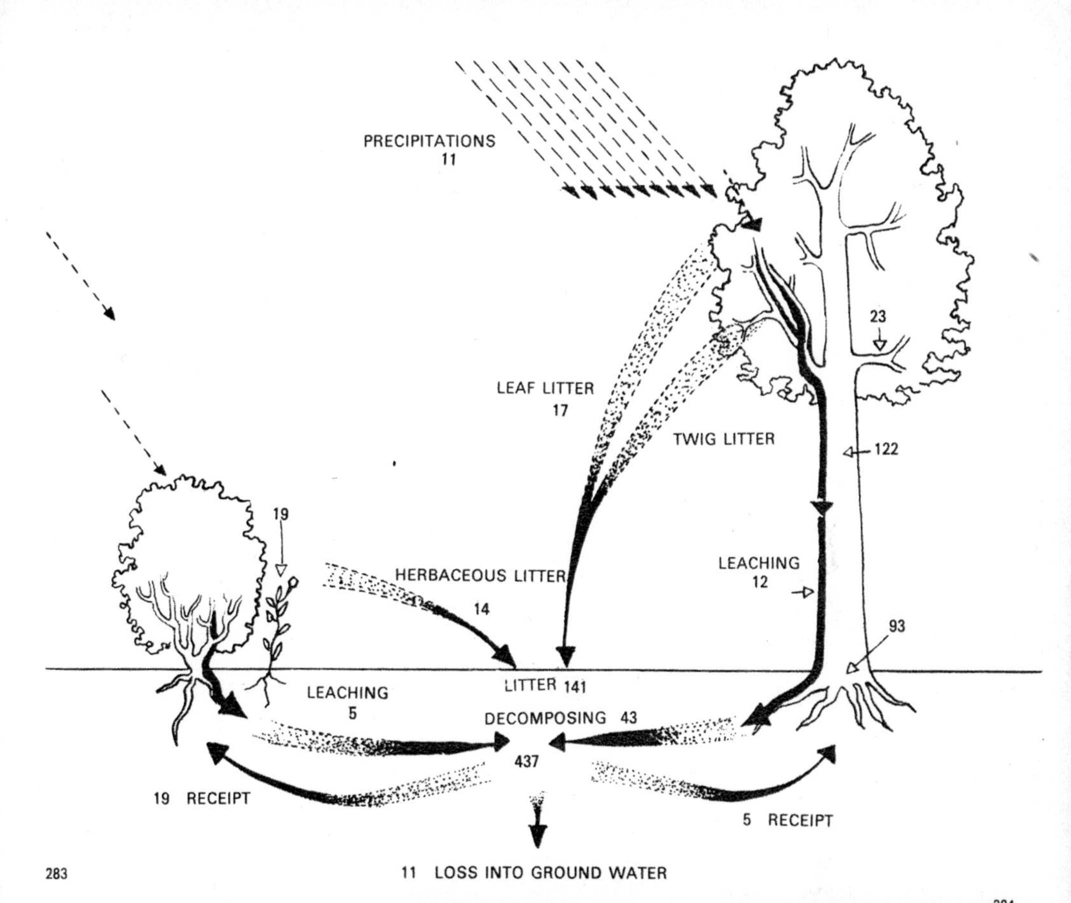
PRECIPITATIONS
11
LEAF LITTER
17
TWIG LITTER
23
122
19
HERBACEOUS LITTER
14
LEACHING
12
93
LITTER 141
LEACHING
5
DECOMPOSING 43
437
19 RECEIPT
5 RECEIPT
11 LOSS INTO GROUND WATER

283

284

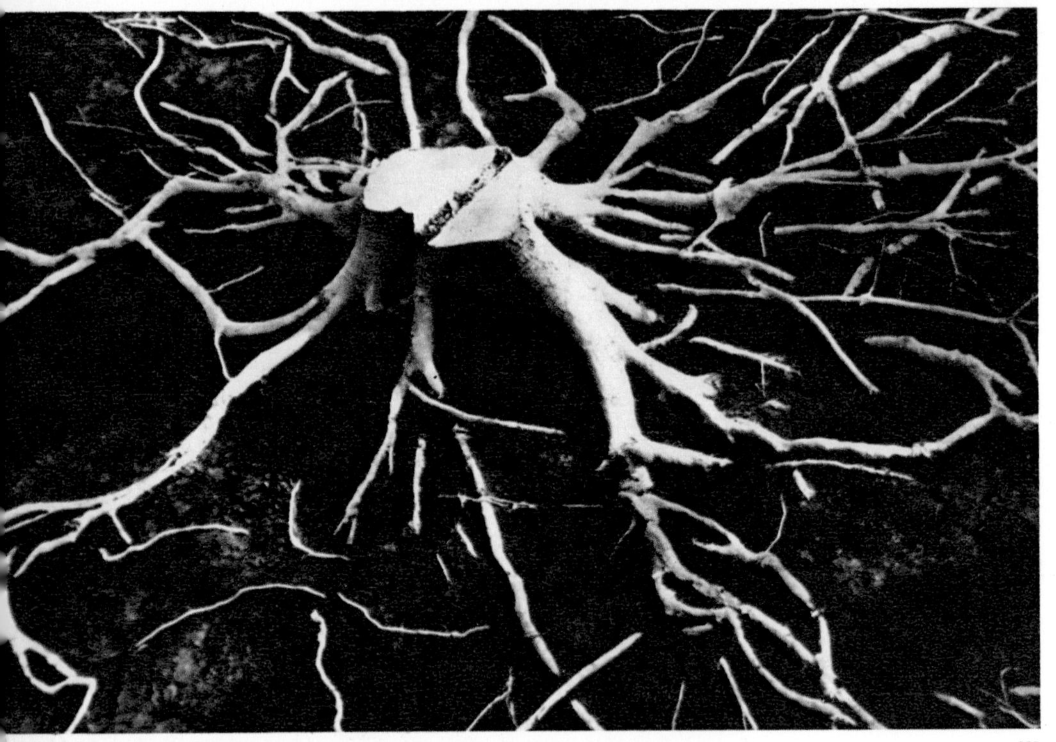

285

286

285 Most forest tree roots are located near the surface of the soil where vital nutrients are released from the humus; Common Beech *(Fagus sylvatica)*, painted white.

286 The soil surface in the tropical rainforest is full of active roots that rapidly absorb nutrients released from the forest litter.

when it is decomposed are they returned to the soil where they become available to the roots of trees. The absorptive roots readily take up these released nutrients in the shallow surface layer of soil which is barely five centimetres deep. If the partly decomposed litter on the floor of the tropical forest is scraped aside, a dense network of competitive roots is revealed (Fig. 286). Some species of fungi participate in the root nourishment of trees. They make a mass or network of thread-like tubes (mycelium) on or inside the roots, which help release and transport nutrients from the soil to the conducting tissues of the tree. In return, the tree provides the fungus with a source of energy in the form of sugar (Fig. 287).

Parasites and saprophytes

Except for a few parasitic or hemiparasitic trees most woody plants of the forest are autotrophic, that is, they are able to synthesize their own nutrients. Flowering plants, however, also include heterotrophic forms that are unable to make their own food and are dependent on the autotrophic species for their nourishment. They are parasites, hemiparasites, or saprophytes. Saprophytic plants obtain energy and food via the fungi linked with their roots which themselves obtain nourishment from decaying organic matter. Examples of such saprophytes in coniferous and broadleaf forests are the Coralroot *(Corrallorhiza trifida)* and Bird's Nest *(Neottia nidus-avis)*. In tropical forests saprophytic flowering plants are far more numerous and even form specialized families (e.g. Burmaniaceae).

Most rapacious of the flowering plants are the parasitic plants of the Orobanchaceae family, which includes the widespread Toothwort *(Lathraea squammaria)* and many tropical species of the Balanophoraceae, Hydnoraceae and Rafflesiaceae families. These parasites produce root-like outgrowths called haustoria through which they absorb food and energy from the conducting tissues of the green plants that are their hosts. Such a parasite, of course, is spared the necessity of using energy to build vegetative organs and can concentrate on the production of flowers and fruits. One of the most dramatic examples is given by the genus *Rafflesia* (Fig. 288).

In 1822 botanists the world over were astounded by the description of an extraordinary flowering plant in the rain forest of Sumatra — a plant that consisted only of a giant flower,

287

287 Numerous fungi live on the roots of trees, *Amanita crocea* for example; their mycelium either spreads over the surface or penetrates inside into the root tissues.

288

288 Flower of the parasitic *Rafflesia.*

289 The Rafflesiaceae family is represented by *Cytinus hypocistis* in the Mediterranean region; its flower, however, is much smaller than that of its tropical relatives.

289

290 The tropical *Tapinanthus farmari* of the Loranthaceae family absorbs solar radiation through its green leaves but takes water and nutrients from the host plant through haustoria.

290

without any leaves, protruding from the thin root of the host liana. The new plant, which was named *Rafflesia arnoldii* in honour of Sir Thomas Raffles, the founder of Singapore, fascinated experts by the remarkable size of the flower (almost a metre in diameter) as well as by the outstanding reduction of the vegetative organs (leaves, twigs and roots). In the course of its evolution this parasitic plant had lost all normal vegetative structures and was reduced to a system of haustoria and large reproductive organs.

The flowers of *Rafflesia* are produced by an unusual process — inside the tissues of the host plant and at the same time within the tissues of *Rafflesia* itself. The embryo of the flower bud is formed in the intercellular space of the thin surface layer of *Rafflesia* tissues embedded in the host liana. The bud gradually grows and after penetrating first its own protective cover and later the tissues of the root or stem of the host plant it finally emerges.

The flowers of *Rafflesia* grow from the roots of woody lianas at ground level, as a rule, but they may also grow out the woody stem several metres above the ground. The species is apparently dioecious, which means that individual plants are either male or female.

The flowers are solitary for they have a very slow rate of growth. Observations have shown that the flower takes several years to develop, whereas its full display and reproductive function takes only a few days.

The sophisticated coloration of the huge

291 Huge earthworms of the tropical forest play an important role in the decomposition of humus in forest soils.

292 Springtails (Collembola) crush forest litter.

293 In swampy soils with insufficient oxygen some tropical trees grow breathing roots—pneumorrhizae; *Anthocleista nobilis,* west Africa.

294 Centipedes of the genus *Lithobius* are agile representatives of animal life in the soil.

291

292

293

294

blooms and the pronounced stench, resembling the smell of decaying meat, attract swarms of flies which lay their eggs in the flowers and at the same time carry out the process of pollination.

The fruit of *Rafflesia* is large and fleshy with a roughly furrowed surface. Inside is a labyrinth of irregular chambers and corridors containing seeds which are about one millimetre long and half a millimetre wide. To date it is not known how these seeds are dispersed. Possibly the whole fruits, parts of the fruit, or individual seeds are dispersed by rats, wild boars, ants or termites. Just as little is known about how the seed germinates and how it gets to the host plant.

Comparative studies have shown that related species of *Rafflesia* occur not only in the tropics but also in the Near East, the Mediterranean region (Fig. 289), southern Europe and southern North America.

Less aggressive are the hemiparasites which include the numerous members of the Loranthaceae family (Fig. 290) and many trees and shrubs of the sandalwood (Santalaceae) family. Hemiparasites obtain only part of their nourishment from the host plant, augmenting this with food manufactured by photosynthesis in their own leaves. The Loranthaceae include certain tropical species that derive nourishment from other hemiparasites of the same family. Thus there are always many links in the chain of decomposers in the forest ecosystem. The sandalwood family includes large trees, such as the evergreen *Santalum album* of India and south-east Asia. This partial parasite yields sweet-scented wood used in commerce. The African *Okoubaka aubrevillei* of the same family grows haustoria that penetrate the roots of neighbouring trees eventually killing them. Such a tree is thought by forest dwellers to possess supernatural powers.

295

296

295 The longhorn beetle *Pachyta quadrimaculata* lives on flowers in coniferous forests and its larvae develop in the wood of the spruce.

296 The tropical millipede *(Julus)* grows to a length of 10 cm.

The larva of the Poplar Admiral *(Limenitis populi)* looks like a prehistoric monster; in the forest ecosystem it is merely an unimportant link in the food chain.

The fruit of the Raspberry *(Rubus idaeus)* contains concentrated energy in the form of sugars which are sucked up by the Common Wasp *(Vespa vulgaris)*.

The Koala *(Phascolarctos cinereus)* feeds only on eucalyptus leaves.

The larva of the Pine Sawfly *(Diprion pini)* causes great damage in pine forests.

The Checkered Beetle *(Thanasimus formicarius)* is a predaceous insect which hunts bark beetles.

The fruiting body of a wood-damaging fungus is a sign that an old fir is infested with mycelium.

297 Forest ants take part in the first stage of decomposition on the soil surface.

297

Most forest parasites and saprophytes are fungi and bacteria. Together with micro-organisms that cause decomposition they hasten the cycling of nutrients and the flow of energy. In rare instances, of course, decomposers are unequal to the task and cannot break up the dead organic matter. This is the case in peat forests which cover large areas in North America and Eurasia. The vegetation there is composed mainly of mosses, sedges, heaths and certain specialized pines. In such a biome the work of decomposers is slowed by the lack of nutrients and presence of inhibiting substands in the soil substrate and also by an insufficient supply of oxygen for respiration. Certain trees growing in tropical peat bogs and wetlands solved this problem of insufficient oxygen supply to the roots by developing special breathing roots (Fig. 293).

Herbivores and carnivores

The soil is the home of most decomposers and consumers (herbivores, carnivores, and omnivores, the latter consuming both plants and animals). The untrained eye generally sees in forest humus only earthworms (Lumbricidae) which both in number and weight of biomass are often the dominant animals of the forest. In tropical regions there are earthworms of enormous size (Fig. 291). However, there are hundreds and thousands of other invertebrate species besides earthworms at work on the surface as well as beneath the soil (Fig. 292).

Some soil fauna such as worms, insect larvae and adult insects feed directly on living roots and fallen leaves or seeds. These herbivores are attacked by predatory carnivores such as predaceous beetles and when they die

298

298 The larva of a click beetle (*Elater* sp.) hunts other insect larvae under the bark of a stump.

299 Bees suck the nectar of flowers and collect the pollen of forest plants.

299

both the herbivores and carnivores become food for a great variety of decomposers—protozoa, worms and arthropods. This last group is represented in forest humus by variously adapted spiders, mites, centipedes (Fig. 294), millipedes (Fig. 296), beetles, termites and ants (Fig. 295 and 297). In tropical forests, termites and ants, together with earthworms, turn over large amounts of soil, often favourably affecting its fertility. Termites and ants frequently carry mineral soil up the trunks of trees to the crowns thus helping create hanging gardens with fertile soil in which epiphytes thrive.

The degree to which soil contains animal life, roots and micro-organisms is directly proportional to its fertility, as a rule. Because all soil organisms must breathe, soil biologists measure fertility by determining the amount of carbon dioxide released by respiration in a certain volume of soil. Such 'biological activity of soil' is highest in the humus-rich soils of broadleaf forests.

Herbivores, carnivores and omnivores are found above ground wherever there is an available supply of food, where there are no poisonous or repellent substances and where the territory is unoccupied. These may be monophagous (feeding on only one kind of food) or polyphagous (eating many kinds of plants and animals). Caterpillars that feed only on a particular broadleaf tree or conifer are monophagous as is the Koala *(Phascolarctos cinereus)* which feeds exclusively on eucalyptus leaves.

Forest herbivores consume a great part of

300

301

302

300 The leaf beetle *Chrysochloa cacaliae.*

301 The Click Beetle *(Elater sanguineus).*

302 A Purple Emperor *(Apatura iris)* sucks the sap of a damaged oak.

303 A herd of female Red Deer *(Cervus elaphus)* in a mixed broadleaved forest.

304 The largest herbivore of northern coniferous forests is the Elk *(Alces alces);* Alaska.

305 The Fallow Deer *(Dama dama)* is becoming established in European forests to which it was introduced.

304

305

the plant biomass. They are adapted for feeding not only on certain species of plants but also on specific parts such as roots, wood, seeds, fruits, leaves, nectar and pollen (Fig. 299, 300 and 302). The greatest amounts are consumed by herbivorous vertebrates, mainly birds and mammals. In various parts of the world, forests harbour large bands, herds or groups of koalas, rodents, ungulates, lemurs and monkeys. Their grazing markedly affects the structure and composition of the forest, particularly in the case of ruminant ungulates such as the Elk *(Alces alces)*, several kinds of which are still found in the forests of Alaska, Canada and Eurasia (Fig. 304). A full-grown male elk weighs up to 600 kilograms and when grazing nibbles the branches and bark of trees and shrubs. It also feeds on aquatic plants in the neighbourhood of the forest.

A more common ungulate of the temperate forests of Eurasia and North America is the Red Deer *(Cervus elaphus)*. It is highly prized as game and thus maintained even in regions with developed forestry (Fig. 303 and 308). Red deer normally graze on herbaceous plants but when snow covers the ground they also

306

306 The coloration of the Two-toed Sloth *(Choloepus didactylus)* is a mosaic of dark and pale colours which blends with the pattern of light and shade in the treetops.

nibble at young trees thus making natural or planned regeneration of the forest difficult or even impossible. When food is in short supply deer even peel the bark from trees, making them vulnerable to fungal attack.

At one time the life of the forest was affected by the bovines (Bovidae), represented in Europe by the Bison *(Bison bonasus)* (Fig. 309) and found today only in a few reservations and game preserves. In North America there are still surviving remnants of once huge herds of Buffalo *(Bison bison)* which can greatly affect the balance of the forest ecosystem when they range through the countryside. In the tropical forests of Africa large herbivorous ungulates include the Lesser Kudu *(Strepsiceros imberbis)*, Bongo *(Boocercus euryceros)*, Black Buffalo *(Syncerus caffer nanus)* and the rare Okapi *(Okapia johnstoni)*. Herds of African Elephants *(Loxodonta africana)* and Indian Elephants *(Elephas maximus)* (Fig. 307) also have a marked effect on the structure of forests.

An example of how even a very inconspi-

307

307 A mammal as large as the Indian Elephant *(Elephas maximus)* consumes large quantities of food and is a destructive element in the forest.

308

308 In forest reserves where food is scarce the Red Deer *(Cervus elaphus)* peels the bark from forest trees in winter.

309 The Bison *(Bison bonasus)* was once a common herbivore in European forests.

309

310

311

310 Molluscs graze in the forest at their own slow pace; Edible Snail *(Helix pomatia)*.

311 The Cockchafer *(Melolontha melolontha)* feeds on the leaves of European trees, sometimes stripping them bare.

cuous animal has its place in the forest ecosystem is shown by investigations into the influence of sloths such as *Choloepus didactylus* (Fig. 306) and *Bradypus tridactylus* on energy flow in the tropical rainforest.

The Three-toed Sloth *(Bradypus tridactylus)* is a herbivore that spends most of its time in the treetops, descending to the ground to defecate and when it decides to move to another location. A single sloth covers an average of thirty-eight metres a day, moving mostly at heights of more than ten metres above the ground and inhabiting a territory of about one and a half hectares. During the course of observations covering a period of six months, a three-toed sloth visited on average forty-one trees, embracing twenty-four different species. Each individual had its favourite tree which it visited more frequently. Samples of stomach content showed that sloths fed on the leaves of more than thirty species of tropical trees. These animals, therefore, do not feed only on the leaves of cecropias as was previously assumed.

Because it was impossible to observe sloths in the act of feeding or to estimate the quantity of food consumed high up in the treetops, scientists determined the frequency of their defecation, the amount of excrement, and the efficiency of their digestion. They found that, on average, a sloth's daily consumption is fifteen grams of fresh foliage per day which, in relation to body weight, is five grams of foliage per kilogram per day. This is only about a quarter of the amount consumed by other mammals in tropical forests. Determining the number of sloths on the given territory was also done by an indirect method, counting the number of fecal heaps. This yielded the information that the average number of sloths was between eight and nine per hectare.

The total number of sloths in the forest far exceeds previous estimates and gives cause for considering the important influence of sloths on the forest ecosystem.

Sloths have a low basic metabolism (about half that of other mammals) and are thus able to lower their body temperature when the ambient temperature drops at night, thus limiting their consumption of food. This enables large populations to exist on a fairly limited territory and at the same time permits a secret and sedentary mode of life that does not force sloths to move about rapidly. Their life strategy is, to say the least, original, but ecologically comprehensible.

312

312 The Striped Fieldmouse *(Apodemus agrarius)* is an important consumer of forest seeds and fruits.

313 The Red Squirrel *(Tamiasciurus hudsonicus)* is an omnivore but the forest ecosystem is affected most by its collecting of spruce cones; Alaska.

313

314

314 The Canadian Beaver *(Castor canadensis)* affects the ecology of forests by its felling of trees and building of dams.

315 Long-tailed Tits *(Aegithalos caudatus)* changing guard at the nest.

315

Many species of mammals, and particularly rodents, are well adapted to feeding on hard seeds, juicy fruits, buds, bark, wood and young shoots. A great many rodents are omnivorous. Common in the coniferous taigas of North America is the Red Squirrel *(Tamiasciurus hudsonianus)* (Fig. 313) which is especially fond of the seeds of the white spruce, but during the course of the year also eats meat. The size of the red squirrel population is directly proportional to the harvest of white spruce seeds in a particular year. Tree squirrels and flying squirrels are ideally adapted to life in the treetops. The Siberian Chipmunk or Borunduki *(Eutamias sibiricus)* and the Eastern Chipmunk *(Tamias striatus)* of the forests of eastern Canada and the United States are also omnivores. Important forest inhabitants in all regions are fieldmice, such as the Yellow-necked Fieldmouse *(Apodemus flavicollis)*, and the Striped Fieldmouse *(Apodemus agrarius)* which inhabits forest margins (Fig. 312). Various species of rats are also avid consumers of fruits and seeds in the forest ecosystem.

Quite unusual is the role played by beavers in the ecology of the forest (Fig. 314). The European Beaver *(Castor fiber)* has become very rare but the common Canadian Beaver *(Castor canadensis)* is an important factor in North American forests, not only on account

316 The Wood Pigeon *(Columba palumbus)* usually feeds on seeds but occasionally eats earthworms and molluscs.

317 Flocks of African Grey Parrots *(Psittacus erithacus)* attack trees laden with ripe fruits and seeds.

316

317

318

318 A Wryneck *(Jynx torquilla)* with its beak full of insects.

of its 'logging' activity but also because it constructs dams across streams in valleys. The resulting lakes and the waterlogged soil behind the dam may cause the destruction of the original forest over an area of several hectares.

In respect of consumption of plant material within the forest, mammals may be surpassed by overpopulations of insects. Caterpillars, like those of the Black Arches Moth *(Lymantria monacha)*, Gypsy Moth *(Lymantria dispar)* or Green Tortrix Moth *(Tortrix viridana)*, may cause serious damage by divesting trees of all their leaves and thus giving rise to restricted growth or even the death of the trees. Grubs of the Cockchafer *(Melolontha melolontha)* feed on and damage the roots of plants and about every four years adult beetles feed on the leaves of forest trees (Fig. 311). The rate at which herbivorous molluscs feed on leaves and young shoots is slow and the damage caused is negligible compared with that caused by caterpillars. The Garden Snail *(Helix pomatia)* is one such mollusc commonly found on forest margins (Fig. 310).

Herbivorous birds have no difficulty in moving among the forest trees. Most frequently found are pigeons (Columbiformes), parrots (Psittaciformes), game birds (Galliformes) and songbirds (Passeriformes). Flocks of Wood Pigeons or Ring Doves *(Columba palumbus)* and Stock Doves *(Columba oenas)* consume acorns and other forest fruits in Europe's forests (Fig. 316). Sadly the Passenger Pigeon *(Ectopistes migratorius)*, which was a common inhabitant of North America's forests, with flocks numbering millions, is now extinct. The last wild specimen was shot in 1909 and the last one in captivity died in 1914 in the Cincinnati Zoo.

Large and noisy flocks of parrots are consumers of plant food in subtropical and tropical forests. When the fruits of a particular species of tree are ripening the parrots remain in one section of the forest until they have eaten all the fruits there. One example is the African Grey Parrot *(Psittacus erithacus)* which moves from one fruit-laden tree to another in the rain forests of Africa (Fig. 317).

Game birds move about both in trees and on the ground, but they are not strictly vegetarians. In European forests the Capercaillie *(Tetrao urogallus)* pecks out the buds of trees and eats the leaves, but it also takes ants and other small insects on the ground (Fig. 320). In view of the large territory occupied by a single pair

319

319 A Sparrowhawk *(Accipiter nisus)* on the nest.

320 The Capercaillie *(Tetrao urogallus)*.

320

321

321 The Canadian Grouse *(Canachites canadensis)* plucks the buds from trees and gathers insects in the northern forests.

322 A Hawk Owl *(Surnia ulula)* with a Weasel *(Mustela rixosa)* in an Alaskan forest.

323 The Robin *(Erithacus rubecula)* feeds a young Cuckoo *(Cuculus canorus)* in its nest.

322

324

of these birds, however, they cannot be said to have a marked effect on the forest ecosystem. The same is true of the Hazel Hen *(Tetrastes bonasia)* and its North American relative *Canachites canadensis* (Fig. 321), which feed on berries, the buds of trees and small animal life within the soil.

Switching from plant to animal food and vice versa is not at all unusual among forest animals. In some species such a change is simply a matter of necessity when one or other source of food is in short supply. Other species simply eat everything that is juicy and edible. Not at all particular in its choice of food is the Wild Boar *(Sus scrofa),* which roams through the forest and eats whatever it comes across be it roots, tubers, mushrooms, tree fruits, worms, insect larvae, adult insects, small amphibians, rodents or even small birds (Fig. 326). The same is true of bears (Ursidae), whose bloodthirstiness is usually much exaggerated. For instance the Brown Bear *(Ursus arctos),* found to this day in the larger forests of Eurasia, feeds on berries, nuts, buds, the honey of wild bees and of course molluscs, insects, reptiles, birds and mammals as well as carrion. Likewise the North American Grizzly Bear *(Ursus arctos horribilis)* is primarily

325

324 Feline beasts of prey live in all types of forest; the European Lynx *(Lynx lynx).*

325 The northern forests of Eurasia and America are the home of the Wolf *(Canis lupus)* which hunts ungulates in winter.

Colonies of ants play an important part in the forest ecosystem throughout the world.

The Northern Birch Mouse *(Sicista betulina)* of northern Europe.

a vegetarian despite its frightening name. Only occasionally does it dig out ground squirrels or devour the young of forest ungulates and the carcasses of dead animals. (Fig. 327)

At all levels of the forest ecosystem there are animals which are primarily carnivorous. Best adapted of these are beasts of prey, which formed, and in many instances still form, the top level of the food pyramid of forest producers and consumers. Members of the cat family (Felidae), and in particular the large cats, have no natural enemies in the wild other than man, who from fear, or for the love of the hunt, or to win a trophy or precious skin, or even for the well-intentioned purpose of preserving other game, has decimated whole populations of these magnificent beasts. In North America the chief representative is the Puma or Mountain Lion *(Puma concolor)*, which is an expert climber, leaps great distances and easily downs even large ungulates. In the forests of South America it is the Jaguar *(Panthera onca)*, which moves skilfully among the treetops. It is an excellent swimmer, which is a great advantage in country interlaced with large rivers.

Most widely distributed of the large felines is the Leopard or Panther *(Panthera pardus)*, the most greatly feared beast of Africa's for-

326

326 The Wild Boar *(Sus scrofa)* is a typical omnivore.

327 The North American Grizzly Bear *(Ursus arctos horribilis)* eats only the berries of forest plants for the greater part of the year.

327

328

ests and also a well-known inhabitant of the forests of southern Asia. Occasionally it is even found as far north as the northern taiga. Acknowledged king of the jungles of southern Asia and the Far East is the Tiger *(Panthera tigris),* which is excellently adapted for hunting in the forest but is unable to climb trees. The forests of Europe have not had a really large cat since the end of the last glacial epoch but this niche has been taken over with success by the European Lynx *(Lynx lynx),* which successfully hunts even large ungulates (Fig. 324).

A widely distributed representative of the canine family (Canidae) in the forests of northern Eurasia and North America is the Wolf *(Canis lupus)* (Fig. 325). In winter, packs of wolves hunt forest ungulates, whereas in summer they are content with smaller animals or even a diet of herbaceous plants. The same area is the home of the Red Fox *(Vulpes vulpes),* which feeds mostly on small rodents, and in a lesser degree on hares, rabbits, capercaillies, hazel hens and pheasants. Much the same fare is eaten by the weasel family (Mustelidae), the Glutton or Wolverine *(Gulo gulo)* being the only member that dares hunt smaller ungulates like roedeer, or the young of red deer, elks and reindeer in the coniferous taigas.

Beasts of prey are of importance in the for-

329

328 The Garden Spider *(Araneus diadematus)* captures prey in its spreading web in the twilight of the forest.

329 Ground beetles (Carabidae) feed on animal material.

est ecosystem because they regulate the numbers of herbivores. In territories inhabited by the larger predatory mammals there is no danger of forest ungulates multiplying to excessive proportions or congregating in large numbers in certain spots. Predacious animals also serve as scavengers, removing weak and diseased individuals. Man, who has driven out beasts of prey from cultivated forests, has taken upon himself the responsibility of fulfilling the top level role in the food pyramid. His performance, however, is not satisfactory because as a hunter he shoots animals according to criteria other than those required to maintain the balance of the entire ecosystem. The result is the overpopulation of ungulates, damage to woodland by nibbling and peeling, and the impossibility of natural regeneration of the forest.

Insectivores such as shrews (Soricidae) feed mostly on insects, but they also consume worms, molluscs, frogs and small birds; plants also form a part of their diet.

There are also flesh-eating species of birds adapted to feed on insects and vertebrates. Various woodpeckers skilfully climb up tree-trunks and chip both bark and wood with their beaks as they look for insects. Hornbills (Bucerotidae) capture somewhat larger prey in the tropical forests with their strong bills. The balance of the forest ecosystem would soon be upset if the insects living there were not devoured by the ubiquitous thrushes (Turdidae), tits (Paridae), nuthatches (Sittidae), treecreepers (Certhiidae), and crows (Corvidae). Crows, of course, also hunt small vertebrates.

Birds of prey (Falconiformes) and owls (Strigiformes) are greatly specialized for hunting live prey. In temperate and tropical forests their members include various species of kites, hawks, vultures, eagles, buzzards, falcons and kestrels. The Eurasian Eagle Owl *(Bubo bubo),* is particularly worthy of note for its diet includes large mammals and birds. Small owls feed on commensurately smaller prey, chiefly small rodents. Nonetheless, the Hawk Owl *(Surnia ulula)* may capture even a weasel (Fig. 322). Here we have a case of one top-level carnivore devouring another top-level carnivore for both are in the top rank of the food pyramid within the forest ecosystem.

The arthropods (Arthropoda) include many groups and species which are predaceous. For example, ground beetles (Carabidae) (Fig. 329) feed on other insects and their larvae, as

330

330 The ant *Camponotus ligniperda* chews corridors in wood but hunts its food outside the nest.

331 *Myelophilus piniperda* lays its eggs under the bark of a pine.

331

332

332 Forest trees are constantly in danger of attack by parasitic fungi; the Honey Fungus *(Armillaria mellea).*

333

333 The bracket fungus *(Fomes fomentarius)* is a common parasite of weakened trunks of the Common Beech *(Fagus sylvatica).*

334 Dead oak trunks and stumps are decomposed by the saprophytic polypore *Daedalea quercina*.

335 A beech trunk riddled by fungi falls to the ground.

334

335

336

336 The instant a tree falls to the ground decaying processes are speeded up.

well as on worms and small molluscs. Another predaceous arthropod is the Garden Spider *(Araneus diadematus)* which catches prey in its web in the forest undergrowth (Fig. 328).

The effect of the trees

Most important for the flow of energy and cycling of nutrients in the forest ecosystem is the developmental cycle of the prevailing trees which play the dominant role in the life of the forest community. The germination of a new seedling and its successful growth adds to the community a new member which absorbs energy and food, at the same time becoming a source of both. The sudden death of a mature tree is equally important in the life of the forest for immediately the ecosystem loses a link on which many further links were dependent. The result is a sort of ecological vacuum which sooner or later is filled by another tree.

Apart from catastrophes such as gales or fire which cause immediate death, a tree does not die all of a sudden. In a very old tree the physiological functions are slowed and the weakened organism is attacked by parasitic fungi and parasitic insects (Fig. 330 and 331). These parasites consume the host's store of food, destroy its tissues and prepare the way for yet further parasitic attack. A very common parasite of forest trees is the Honey Fungus *(Armillaria mellea)* which attacks any weakened tree, not only damaging it but also reducing its resistance to further infection (Fig. 332). The damaged trunks of old beeches are rapidly infested by the bracket fungus *Fomes fomentarius* which causes the trunk to break and in time fall (Fig. 333 and 335). The boles of dead beeches, hornbeams and oaks are infested by the mycelium of the related *Daedalea quercina* which disintegrates the compact wood and prepares the way for other parasites and saprophytes — bacteria, protozoans and insects.

Decomposition of dead forest trees takes decades, and in the cold tundras of the north and in high mountains it may take several hundreds of years. Old stumps and fallen tree trunks (Fig. 336) are gradually covered by vegetation beneath which they crack, decay and become dispersed by rainwater. Only roots of dense and resistant wood remain for long in the midst of the forest (Fig. 337). Cracks and decayed or hollowed-out cavities are inhabited by whole populations of saprophytes and parasites. Insect larvae that feed on dead wood are in turn fed on by other parasitic insects. Cavities serve as nests for birds, themselves infested by numerous internal and external parasites. Larger cavities are inhabited by rodents, insectivores and small beasts of prey. A single tree can thus become a miniature zoo.

Sooner or later all the energy stored in the wood of a dead tree either finds its way into other organisms or is returned to the soil. Likewise all nutrients are gradually absorbed by bacteria, fungi, insects and other animals and eventually returned to the soil through their excreta or their dead and decomposed bodies. The resulting humus then serves as a source of nourishment for a new generation of trees and forest herbs.

337

337 Some trees die standing up; mountain spruce in the Tatras, Czechoslovakia.

A COMMUNITY OF GIANTS

The function of trees

Although it is wrong to view the forest merely as a collection of trees, it would also be unjust not to acknowledge the leading role trees play in the forest ecosystem. They are giants not only in size but also by their influence on the atmosphere, soil and other forest organisms. Their growth and production of vast quantities of wood and litter covering the forest floor are important to the existence of populations of mosses, lichens, herbs, birds, mammals, insects, worms and protozoans. It may be possible to imagine a forest which has, for example, no epiphytic algae or lichens or bromeliads. Many forms of animal life, though they consume plants or herbivores, are not strictly necessary to the life of the forest. But a forest would not be a forest without trees, because it would lack the driving force behind the energy flow and cycling of nutrients in its ecosystem.

Trees fulfil the function of builders in the forest because in the process of natural selection during the lengthy course of evolution they have developed long-lived bodies that occupy a large space both above the ground and in the soil. The trunk of a tree has many physical and chemical qualities which modern technology has tried in vain to duplicate. Trees are able to make efficient use of, and also store solar energy, which is available in only limited quantities on certain part of the surface of the earth.

The structure of trees

The roots

The root system of trees, as in other plants, is concealed mostly in the ground, but much is known about it because it is often exposed to view when trees are uprooted by the wind (Fig. 340) or when the soil is eroded by water (Fig. 341) or wind, and when it is removed during scientific excavations. On the banks of watercourses exposed roots continue to grow with success (Fig. 339). Concealed from view in soil or water, roots branch out to form a system of progressively smaller organs (Fig. 342 and 343) and at the absorptive terminal regions new growth is continually taking place to renew damaged tissues. Nutrient solutions in the soil are carried from the roots to the crown through a complex of conducting tissues (vascular bundles) (Fig. 346). There are four types of root systems:

1) a single strong main root (taproot) growing almost vertically downwards and from which small lateral roots spread out at all levels;
2) roots of more or less equal thickness (with smaller lateral roots) radiating from the base of the trunk in all directions;
3) roots spreading from the base of the trunk near the surface to form a flat plate-like system;
4) ground roots supplemented by adventitious roots arching from the base of the tree and rooting some distance away.

These different forms of root systems are a hereditary characteristic of the given species, naturally undergoing changes under the influence of the environment.

There is considerable variety of form in the terminal rootlets of trees. They are often divided into main leading roots and smaller laterals which die after a time. There are various other types of specialized roots: some trees form roots called haustoria with which they invade neighbouring plants and through which they absorb water and minerals from them; sometimes terminal roots are transformed into respiratory organs that facilitate the exchange of gases between the atmosphere and submerged roots; the roots of many trees produce nodules inhabited by bacteria or fungi which fix atmospheric nitrogen (that is combine it into the form of simple compounds of nitrogen, which pass to the tree and finally to the surrounding soil.

Associations between bacteria or fungi and

338 A giant oak

339 The roots of the Common Alder *(Alnus glutinosa)* support a high river bank.

the roots of trees are frequently encountered. In natural forests young emerging roots are soon infested by soil bacteria and fungi present in every forest soil. Some of these bacteria and fungi may be injurious to the trees and thus are troublesome parasites. Some may live on the surface and inside the roots in a mutually beneficial or symbiotic relationship. Some species of bacteria that live on the surface of roots transfer released nutrients to the tissues of the root, while others feed on the secretions appearing on the surface of living tree roots. Most remarkable, however, is the symbiotic relationship between tree roots and fungi. The tree tissues and those of the fungus form a compound structure, or mycorrhiza, in which both organisms are structurally linked.

Mycorrhizae are of two chief types: 1) ectomycorrhizae, in which the fungus forms a mantle over the surface of the roots and the hyphae (individual threads of fungal tissue) extend only into the intercellular spaces in the superficial zones of the root, and 2) endomycorrhizae, in which the hyphae of the fungus have forced their way into and inhabited the actual cells of the root. Only certain types of

339

340 Main parts of the root system of forest trees: a—root spurs, b_1—vertical skeleton roots, b_2—taproot, b_3—horizontal skeleton roots, c_1—terminal short-lived rootlets of limited growth, c_2—continuously-growing terminal rootlets.

341 Exposed roots of a 300-year-old oak.

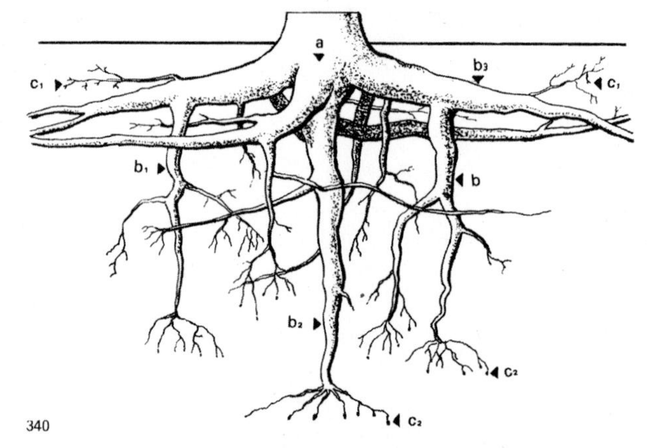

340

341

342

342 Base of the trunk of the tropical tree *Hildegardia barteri* (family Sterculiaceae) with impressive buttresses; Ghana.

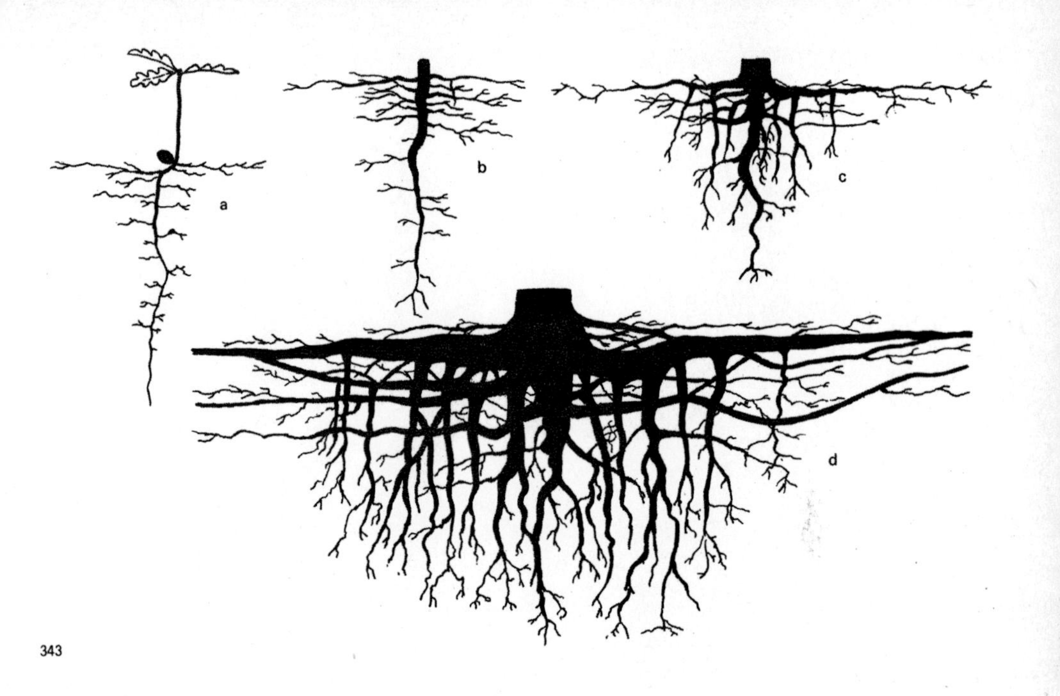

343 Growth of the root system of the Common Oak *(Quercus robur)* in deep loamy soil; a—one-year-old seedling, b—ten-year-old tree, c—25 to 30-year-old tree, d—70-year-old tree.

ectomycorrhizae can be identified with the naked eye—chiefly those that extend mycelia (bundles of hyphae), sometimes strikingly white and at other times black or coloured, outward into the soil (Fig. 345). Some ectomycorrhizae are completely smooth on the surface and look like ordinary smooth roots. Only an anatomical section observed through a microscope reveals the truth. The untrained eye, however, is quite unable to recognize endomycorrhizae, except in a few trees like maple in which they look like slightly swollen nodules or beads.

For a long time all fungi were considered to be parasites. It was only a hundred years ago that botanists and foresters began seriously to consider the mutual benefits between fungi and roots, and it was not until fifty years later that they came up with undeniable proof that mycorrhizae are truly symbiotic—that their association with roots is truly of mutual advantage.

In most plants the process of absorption of water and raw materials from the soil is carried out by root hairs—thin-walled hair-like tubular outgrowths from the active roots. In the case of ectomycorrhizae, the long hyphae which extend outwards into the soil can be thought of as substitute root hairs or even an

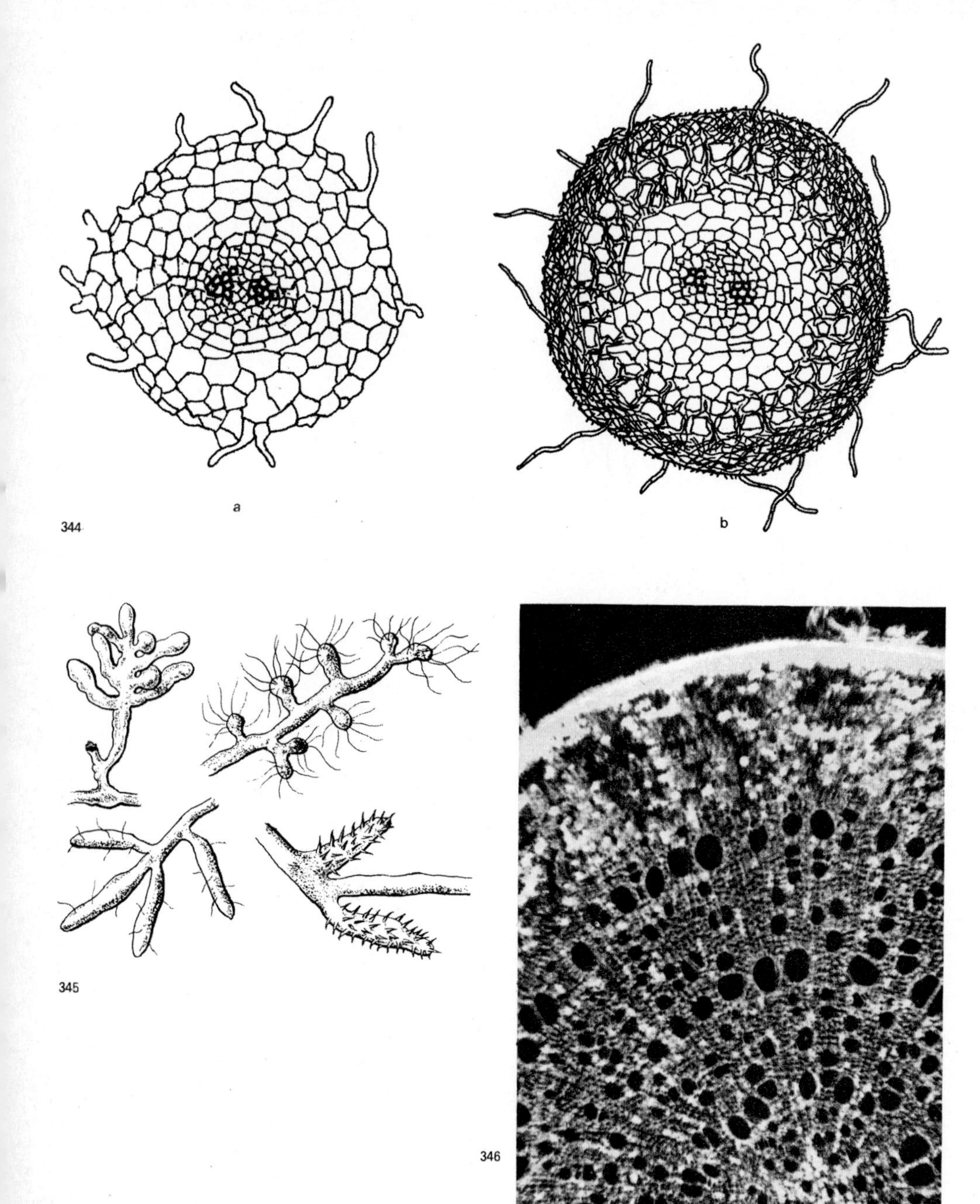

344 Cross-section of (a) a short-lived terminal rootlet of limited growth and (b) of an ectomycorrhiza on a coniferous tree.

345 Various types of ectomycorrhizae of European oaks.

346 Cross-section of the root of a lime tree *(Tilia)* showing vascular bundles.

347

347 Stilt roots of *Xylopia staudtii* (family Annonaceae); Ghana.

348 Stilt roots of a member of the genus *Uapaca* in a virgin forest in Africa.

348

Exposed roots of the giant tree *Chlorophora excelsa* in an African rain forest (the roots are in an African rain forest contrast).

Fossilized stumps of redwoods *(Sequoia* sp.*)* covered by volcanic ash 40 million years ago; Colorado, U.S.A.

Base of the trunk of a Yellow Birch *(Betula alleghaniensis)* with stilt roots growing on a now-decayed stump; Green Mountains, Vermont, U.S.A.

349

349 The curiously shaped trunk and roots of a Norway Spruce *(Picea abies)* which germinated on a fallen tree.

350

350 Adventitious breathing roots (pneumorrhizae) of a *Spondianthus preussii* tree in a flooded forest beside the Tano River, Ghana.

351 Stilted peg roots of *Avicennia germinans* and stilt roots of *Rhizophora racemosa* in mangrove woodlands; Sierra Leone.

improved version as there is so much hyphal tissue in contact with the soil (Fig. 344). Convincing proof that the fungus truly provides the tree with nourishment in this way was obtained by the Swedish botanist Melin and his successors, who, after the Second World War, were able to use radioactive isotopes in their experiments. Similar success crowned later experiments which showed that organic substances are also passed from the trees to the tissues of the fungus. Symbiosis in ectomycorrhizae, then, consists in the fungus providing the tree with nourishment in the form of simple raw materials and the tree providing the fungus with energy-rich carbohydrates, the production of which begins in the green leaves in the crown.

Endomycorrhizae were for a long time believed to be a case of retarded or not unduly injurious parasitism of a tree by a fungus. The seeming absence of contact between the fungus and the external environment gave no ground for assuming mutual benefit. However, detailed observations and experiments proved that these fungi also have their points of entry into the tree roots (in ashes, for example,

351

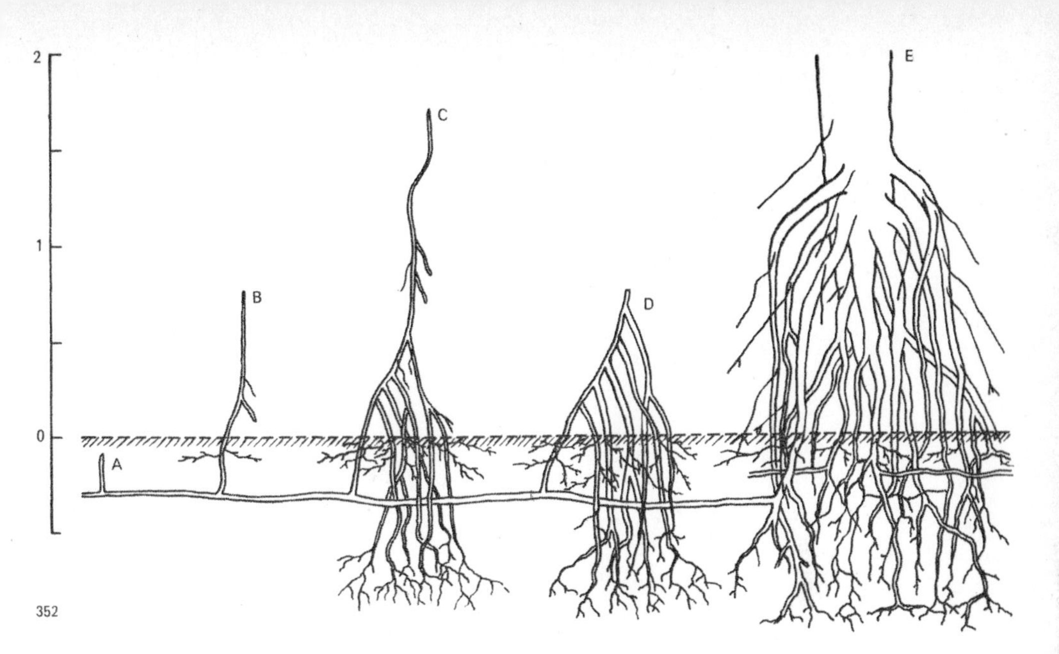

352 Development of the breathing root of the tropical tree *Xylopia staudtii:* A—initial stage with negatively geotropic root, B—advanced stage with lateral roots beginning to grow, C—fully developed breathing root with stilt roots, D—death of breathing root, E—main trunk with stilt roots.

through specialized cells in the epidermis) and that with their growth substances and secretions they greatly influence the growth, development and activity of tree roots.

Extensive experiments by foresters showed that mycorrhizae are important to the growth of trees particularly in unfavourable soils and fluctuating climate. Forest tree species that thrive in extremely poor or dry conditions often have ectomycorrhizae. In the case of tropical trees endomycorrhizae are more prevalent, but the giant trees of some families such as Caesalpiniaceae and Dipterocarpaceae form distinctive ectomycorrhizae together with various species of toadstools.

Roots also have the important function of anchoring the tree in the soil. This may not be easy, particularly in the soft and waterlogged soil of swamps. In tropical forests two different solutions to the problem are found. In the first instance the lower part of the trunk has buttresses and the main roots stand above ground level (Fig. 342). The second solution is an example of the fourth type of root branching. In this instance the trunk of the tree is often of small girth and the main woody substance is distributed in the stout stilt roots (Fig. 347 and 348). Trees in temperate regions have no distinctive buttresses or any kind of stilt roots. This is because these organs greatly increase the surface area of the trunk and in winter may be more easily damaged by frost. In this situation natural selection was ap-

353 Adventitious roots at the base of the trunk of the willow *Salix cinerea* on the banks of a south Bohemian pond.

353

parently against the development of trees with buttresses and stilt roots.

In the yellow birch of America and the Norway spruce of Europe the base of the trunk is sometimes spurred and furnished with organs resembling stilt roots (Fig. 349). The reason in this case is quite different. The two species germinate on old trunks or stumps and their roots grow slowly down into the soil. Later—when the dead trunk has decayed—the upper parts of the normally anchored roots look like the stilt roots of tropical trees. Similarly, the bared roots of alders and oaks cannot be compared with stilt roots.

It has already been said that trees can grow in permanently waterlogged soil either if their roots obtain oxygen for respiration by means of internal chemical reactions as in the case of alders, or if they have special breathing roots that facilitate the exchange of oxygen and carbon dioxide between the surface atmosphere and the submerged roots (Fig. 351). The life of such a tree can also be saved by the growth of adventitious roots which emerge from higher up the trunk (Fig. 350 and 353).

Roots normally grow downwards in re-

354 Trunk of a dead fir in the Jeseník Mountains, Czechoslovakia.

355 (a) The trunk of the Baobab *(Adansonia digitata)* and (b) the Silk Cotton Tree *(Ceiba pentandra)*.

356

357

356 The straight and slender boles of the Large-leaved Lime *(Tilia platyphyllos)*.

357 Trunk of a fir growing by itself in a clearing.

sponse to the force of gravity (they are positively geotropic). This is true of many roots, but, as has already been noted in the case of mangrove woodlands, the roots of some trees grow upwards, and are therefore negatively geotropic. Negative geotropism is characteristic not only of the breathing roots of trees in mangrove woodlands but also of those trees growing in the freshwater swamps of the tropics. Even in the temperate regions of Eurasia the arborescent willow *Salix pentandra* has roots that grow distinctly upwards.

Unexplained, as yet, are the factors that influence the formation of other breathing roots (pneumorrhizae) such as knee roots and stilted peg roots. Knee roots curve upwards towards the surface of the soil at a certain phase of their growth and when they are some centimetres above the ground they suddenly curve back into the soil. The hooked tip of a young knee root at the stage when it curves back toward the soil is a peculiar sight indeed. No less remarkable is the sight of tropical swamps where knees of varying thicknesses protrude above the surface of the soil. Only in some trees *(Bruguiera)* of the mangrove woodlands and freshwater swamps can one particular root grow up and down, in and out of the soil, a number of times. Most species produce only one knee—a branch root growing out of the ground perpendicular to the horizontal parent root and rooting permanently, deep in the soil, following the knee phase in free air.

Stilted peg roots are equally interesting. These roots grow vertically upwards from horizontal roots embedded in the mud. They may reach a height of up to a metre and in the ensuing phase produce lateral roots, all positively geotropic, which grow either straight downwards or at an angle back into the soil. When these have become established and have put out branch roots in the soil they serve as an excellent support for the aerial part of the peg root.

The trunk

The trunk is the most massive part of a tree, being its geometric axis as well as the main pathway of its transport system. In most conifers it is well developed and reaches to the very tip of the slender conical crown. In broadleaved trees it generally divides at one point into thick, upward-growing branches which divide further to form a crown. The trunk of broadleaved trees is usually a smooth cylinder but it may be furrowed or fluted. Such features as the shape of the trunk and the normal maximum height of a tree are genetically determined, but the influence of the environment may cause marked deviations (Fig. 355). There is a considerable difference between, for example, the trunk of a European silver fir growing in the heart of a forest (Fig. 354) and that of a similar tree which has not only grown by itself in a clearing for many years, but has furthermore been damaged by a gale (Fig. 351).

In a closed, broadleaved forest and good soil the Large-leaved Lime *(Tilia platyphyllos)* makes a tall cylindrical bole topped by a small crown (Fig. 356); in parks, tree avenues and as a solitary tree in fields it forms a broad, richly branched crown that begins low down on the trunk; near the upper forest limit in the mountains, the trunk of an old specimen may be twisted completely out of shape (Fig. 358). Likewise there is a great difference between a Scots pine growing in a closed forest and one growing on a dry rock (Fig. 359 and 360) — the former having a slender bole and smaller crown, the latter a thick, prostrate trunk and a large, misshapen crown.

The trunk in great part determines the height of the tree. The tallest of all trees is probably the American Coastal Redwood *(Sequoia sempervirens)*, which can be as much as 110 metres tall. The Australian Mountain Ash *(Eucalyptus regnans)* is sometimes more than ninety metres high. The tallest specimen of *Araucaria hunstenii*, growing in New Guinea, measures eighty-nine metres and in the rainforests of Malaysia there is a specimen of *Koompasia excelsa* which measures eighty-four metres. The coniferous trees of North America and Eurasia generally grow to a height of about fifty metres and include the silver fir, which is the tallest native of Europe. Contrary to popular belief most giant trees of the tropical rainforests grow only to a height of about forty metres (Fig. 361) and are only about a metre in diameter.

The adverse effects of the environment are usually evident both in the shape of the tree and in the height of the trunk (Fig. 362 and 363). However, not all small or low-branching trees are so developed because of adverse environmental conditions. The dwarf habit of many species growing in savanna woodlands and below the canopy of the tropical rainforests is an inherited characteristic. Such trees

358

358 This twisted twig is the main trunk of a Large-leaved Lime *(Tilia platyphyllos)* growing at the upper forest limit in the Giant Mountains, Czechoslovakia.

359

360 Tall, slender boles of Scots Pine *(Pinus sylvestris)* in a dense closed forest.

359 Prostrate trunk and crown with strong but twisted branches of a Scots Pine *(Pinus sylvestris)* growing on rock.

361 The trunks and crowns in the tropical rainforest exhibit a great variety of shapes; Ghana.

grow to a height of only a few metres and the diameter of the trunk does not continue to increase with age, but has a definite limit for each species. The giant sequoias measure up to six metres across at ground level so there is no need to doubt the authenticity of photographs showing a large car driving through a tunnel in the base. The African baobabs are also noted for their girth, in particular *Adansonia digitata.* The trunk of such a tree has even been used as a shop. Even in Europe there are many trees which attain enormous dimensions in old age. One old large-leaved lime in Bohemia was used as a village chapel and held up to thirty people.

The oldest trees are found amongst several species. A Bristlecone Pine *(Pinus aristata)* growing near the alpine forest limit in the Rocky Mountains of America was believed to be 4,000 years old. The South African baobabs are supposed to be even older — 5,000 to 6,000 years. The age of the giant limes and oaks of Europe will probably never be determined with certainty because the annual rings in the centre of the trunk have usually rotted, but they are believed to be between 800 and 1,000 years old. In contrast, even the largest of the broadleaved trees of the tropical rainforests are only between 200 and 300 years old at the most.

A trunk growing as a single stem right from the seedling stage is a typical characteristic of trees. Woody plants with stems branching close to the ground are usually considered to be shrubs. Branching alone, however, is not a reliable distinguishing feature and neither is size, for many a young or dwarf tree is smaller than a robust shrub. Nevertheless, in old age some species of trees form groups of trunks growing from the same base. This often happens when a tree is felled and new stems (coppice shoots) grow from the stump (Fig. 364). Many broadleaved trees (oak, hornbeam, lime) and a few conifers multiply asexually by this method (Fig. 367). Another method of vegetative reproduction is by root suckers which are produced in great numbers by poplars, including the European Aspen *(Populus tremula).* In dry savanna woodlands, which are often destroyed by fire, vegetative reproduction is the determining factor in their survival.

The bark

A protective covering as well as a decorative

360

361

362 Years of grazing and drought have left their mark on the trunk of *Ceratonia siliqua;* Mediterranean region.

363 Common Beech *(Fagus sylvatica)* on former pastureland.

364 Shoots on the stump of a Durmast Oak *(Quercus petraea).*

feature of treetrunks is the variously patterned bark (Fig. 368). It is smooth in those species which have only a single cork-producing layer. In compact bark there are numerous small, round or slit-like patches (lenticels) which serve as pores, permitting the exchange of gases between the atmosphere and the stem tissues (Fig. 366). The harsh effects of the environment often disrupt the compactness of the bark. Unequal freezing and thawing causes longitudinal fissures (frost cracks), which expose the tree to the danger of attack by destructive fungi (Fig. 365). At other times the smooth bark of forest trees is marked by birds which tear pieces out with their beaks. Saddest of all is when smooth bark that has grown for decades is disfigured by the carvings—names and symbols—of man.

Other trees have bark that thickens every year by the addition of a very thin layer of corky tissue. The result is a deeply furrowed rhytidome (Fig. 369) or one that is scaly, the outer scales peeling off successively (Fig. 370). In some species, like the plane, the bark peels off

362

363

364

in large patches or plates, while in many Australian eucalypti, commonly called stringybarks, it comes off in lengthwise strips. Eucalypti in general exhibit great diversity in the structure, ornamentation, colour and fragrance of their bark. The bark of some trees is covered with spines which are either products of the corky tissues or else modified stems or roots. Particularly impressive are the spines of the African *Fagara macrophylla* (Fig. 372).

Many trees can be easily recognized by their bark. The European and American forester can thus identify every tree in the forest without looking at the crown. The same aid to identification is needed by foresters in the tropical regions in their search for commercially important trees amidst the vast wealth of species. Leaves, flowers and fruits, which are the most important botanical features, are of no value for identification purposes in the immense forest stands. In such a situation a good means of identifying a tree is to slash the trunk with a machete (Fig. 373) thus revealing the inner structure of the bark and sapwood underneath. Such a cut exposes the varied layers, coloration, exudations, fragrance and changes in colour caused by exposure to air. Even taste can be an aid to recognition of bark for an experienced forester.

The trunk is made up of true wood called xylem which carries sap up from the roots to the leaves and shoots, produced within a thin cylinder of actively growing tissue called the vascular cambium. Outside this cylinder is a second layer of conducting strands which carry sap from the shoots down to the roots. This layer is called phloem (Fig. 374). The cambium is invisible except in microspic preparations and is best recognized by touch, as the slippery layer revealed when the bark and underlying phloem peel from the wood. The cambium may grow continuously or at seasonal intervals. In the first case, as in trees growing in rainforests, it produces uniform wood that reveals nothing about the tree's age and rate of growth; in the second instance it grows at a varying rate thus producing what appear to be separate layers of wood of different character which look like rings on a cross-section of the trunk. These are called annual rings because they correspond roughly to one year in the life of the tree (Fig. 376). Annual rings make it possible for botanists and foresters to determine the age of a tree, and its rate of growth in girth at various periods of its life. Because the rate of growth depends on the conditions of the environment, annual rings make it possible to determine the climate at various times in the past. An old silver fir may thus portray climatic changes over a period of

365 Bark of beech showing frost cracks.

several centuries while a venerable bristlecone pine provides information about alternating periods of dry and wet years during the course of several millenia.

Detailed measurements of annual rings provide foresters with precise data about the thickness, height and volume of trees of a given species in various sites and at various ages. Selected specimens are cut into metre-long sections, and the number of annual rings are counted and their width measured on each cross-section. On the basis of the decreasing number of rings it is possible to determine the height of the tree at a specific age and according to the respective rings it is possible to determine the girth of the trunk in the various stages of its growth. The results of such stem analyses have been used to compile yield (growth) tables from which it is possible to assess the probable yield of present-day forests in the light of evidence of climatic trends provided by the trees themselves.

The number and width of annual rings can also be measured in live, standing trees. By means of a tool that bores into the trunk of a tree, a core specimen can be taken in which the annual rings from the outside to the centre of the trunk are clearly visible. With such a tool the forester or arborist can examine the rate of tree growth after planting, and can check the results of improvement felling, thinning, or the application of fertilizer.

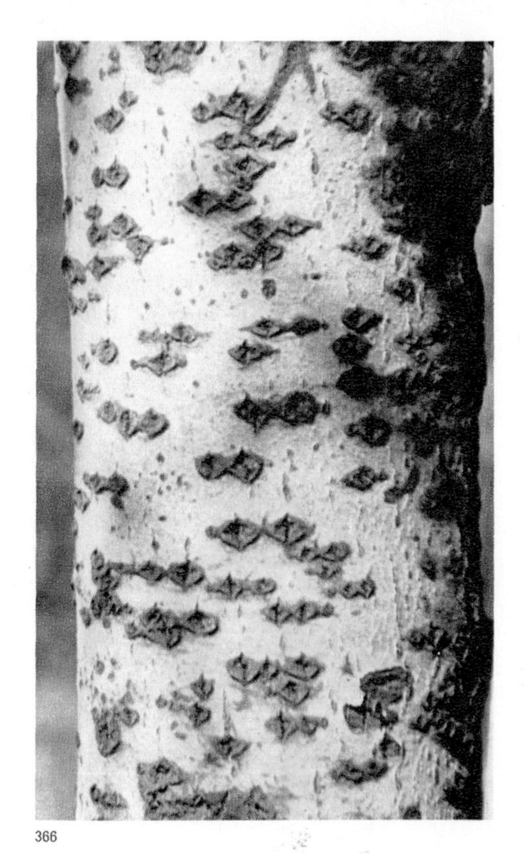

366

367

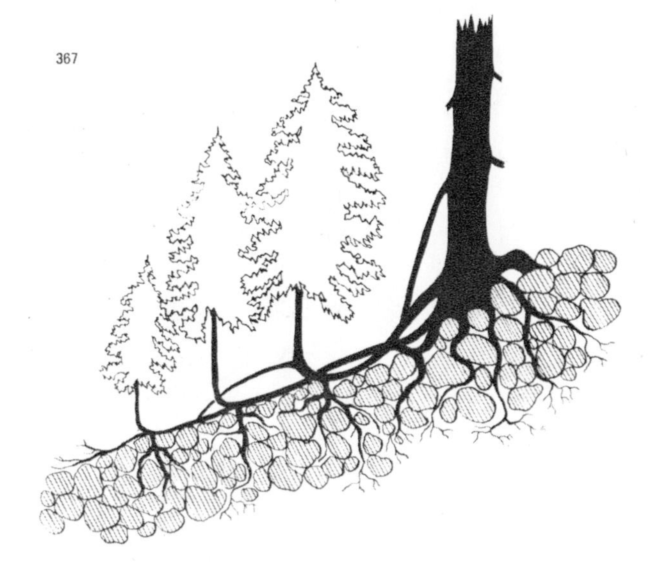

366 Bark of the White Poplar *(Populus alba)* with characteristic lenticels.

367 Vegetative reproduction of the Norway Spruce *(Picea abies)* on stone rubble in a mountainous area.

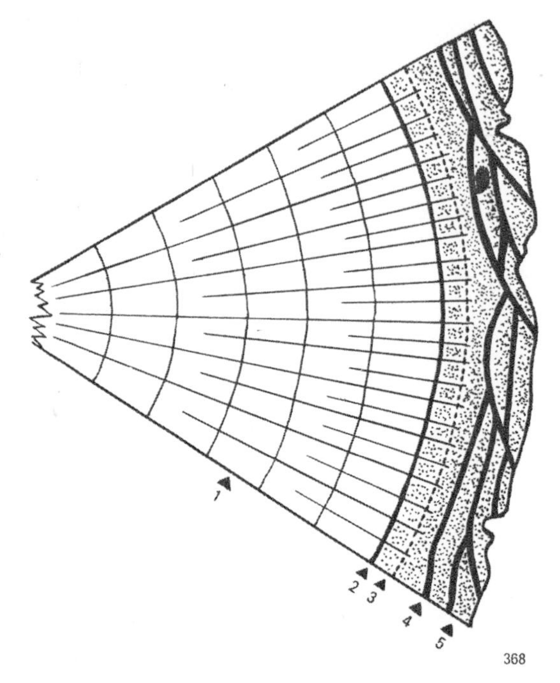

368 The cross-section of a tree trunk: 1 — wood with annual rings, 2 — vascular cambium, 3 — live phloem, 4 — newly forming bark, 5 — protective layer of bark.

369 Furrowed bark of the English Oak *(Quercus robur).*

369

Trunk and aerial roots of the fig *Ficus benjamina,* native to the tropical forests of south east Asia

Part of the trunk of the largest species of eucalyptus—the Australian Mountain Ash *(Eucalyptus regnans);* Tasmania.

370 Scaly bark of the Sycamore *(Acer pseudoplatanus).*

Measurement of the annual rings of standing trees in virgin forests made it possible to reconstruct the entire evolution of the forest as determined by the climate and by the internal ecological relationships resulting from the dying and the natural regeneration of the forest. For instance, these annual rings clearly show when a tree grew in shade and when in light, as well as when the root competition of neighbouring trees was strong and when it was weak. Annual rings readily reveal the harmful effects of industrial air pollutants and the resulting damage to the forest economy is then easily calculated as losses in wood increment.

In older trees the central section of the trunk (heartwood) consists of dead cells. Heartwood does not function as a pipeline for nutrient solutions moving up from the roots to the crown and is usually saturated with impregnating substances. Sooner or later the dead wood is attacked by decomposing fungi and bacteria. The outer layers of wood (sapwood) function throughout the tree's lifetime, serving not only for conducting substances from the roots to the crown but also from the phloem to the inner part of the trunk.

The branches

Both coniferous and broadleaved trees branch in characteristic ways. Some do not branch at all but produce a bunch of leaves at the top of the stem. Many small trees of tropical forests

such as the African *Pycnocoma angustifolia* and *Ficus theophrastoides* exhibit this form which is the prevailing form of monocotyledonous trees. Most palms have a single trunk topped by a crown of leaves and smaller species of Dracaeneae grown as house plants are miniature examples of such trees.

The advantages of the life form of a tree, however, are more evident in species with a branching crown. The tree can thus spread its important food producing and reproductive organs at various levels above the ground. The arrangement of the crown is determined partly by heredity and, particularly in the case of older trees, by the influence of the environment. A solitary Norway spruce growing in the open has a conical crown branching from the ground, whereas one growing in a dense forest loses the lower branches and has only a small crown high up at the top (Fig. 377). In tropical broadleaved trees the crown may be formed by the forking of branches as in *Anthocleista nobilis,* or by the production of lateral branches of varied length, longevity and arrangement in space. The result is a layered, umbrella-shaped or spherical crown (Fig. 379, 381 and 382).

The structure of the crown of broadleaved trees is very intricate in old age, and only the experienced eye can identify the various species in a closed forest solely by the mode of branching. Where a tree is not exposed to the fierce competition of neighbouring trees it may form a characteristic crown that is readily identified at a glance even at a considerable distance. Even a person with a cursory knowledge of American trees can identify the American elm at a distance, in the same way that anyone who has ever been to the savanna woodlands of Africa can identify *Daniellia oliveri* of the Caesalpiniaceae family, with its

372

373

371 Scaly bark of Norway Spruce *(Picea abies).*

372 Spines on the bark of *Fagara macrophylla* in a tropical forest.

373 Slashed trunk of the tree *Celtis mildbraedii* showing the characteristic layers of bark which serve to identify the species.

374 Distinct annual rings are very evident on this decaying oak stump.

375 Bark of the American Sugar Maple *(Acer saccharinum)* peeling off in plates.

376 A cross-section viewed through a microscope shows that a single annual ring is composed of thin spring wood (dark section) and dense summer wood (light section).

crown shaped like an inverted triangle. On closer examination it is possible to distinguish the different arrangements of the branches and their position in space, the flexibility or rigidity of the youngest twigs and the appearance of new buds, the density of the foliage and the colour of the bark. The effects of extreme climatic and soil factors may cause marked deformation of the normal crown. Thus, the beech with slender trunk and broad rounded crown that is common in closed forests is replaced on summits exposed to strong winds, frost and drought by crooked and gnarled specimens (Fig. 383).

The complex structure of the crown is part of the strategy of how to get the utmost benefit from the arrangement of the leaves, flowers and fruits. In some instances there may not even be any leaves. In the Euphorbias of the savanna woodlands, e.g. *Euphorbia candelabrum,* there are no leaf blades and the function of photosynthesis is taken over by the suc-

374

375

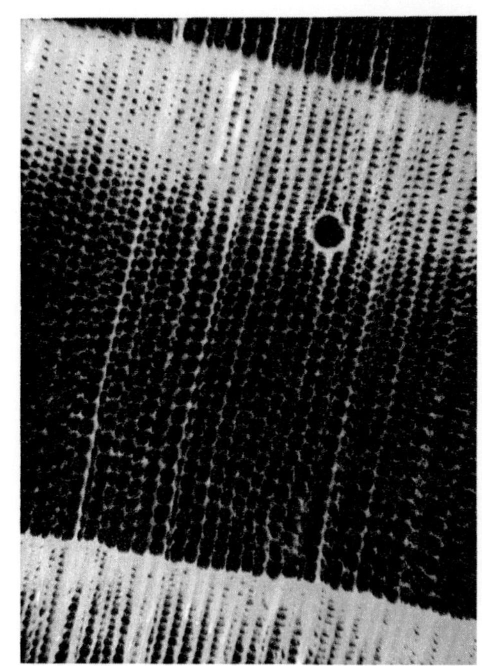
376

culent stems. The result is a rather unusual tree, but a tree nevertheless.

Young twigs, leaves and flowers develop from buds. The buds of trees in temperate regions are already well-developed by the end of the preceding growth period and generally survive the harsh conditions of winter protected by scales. The number and shape of these bud scales, as well as the general shape, size and arrangement of the buds on the twig are characteristic for each species of forest tree (Fig. 384 and 385). For the keen observer buds are a reliable distinguishing feature which make the identification of trees possible even in winter when they are without leaves and flowers. In some instances further aids may be the colour of young twigs, whether or not they are hairy (pubescent), the number and size of lenticels or the shape of leaf scars. It is also possible to cut a twig lengthwise and take note of the wood and pith. The chambered pith of walnuts is an infallible characteristic of the entire genus.

Unlike their temperate counterparts, few tropical trees have buds with scales. Whereas in temperate regions shoots grow during the summer months but remain dormant in winter, tropical trees exhibit marked diversity in the development of new shoots. Some put out new shoots continuously, while in others, periods of active growth alternate with periods of rest several times a year. In many species growth is interrupted once a year for reasons having nothing to do with the environment, and in still others the shoots grow haphazardly in differing succession on each separate twig. In the tropical jungle the life of trees is not synchronized in the same way as in the forests of Europe and North America. At any time of the year tropical trees can be found in various stages of development, and at any one time, in the crown of a single tree, some twigs may be in flower, some may be putting out leaves, some may be quite bare while others are covered with ripe fruit. One such example is the Silk Cotton Tree *(Ceiba pentandra).*

The leaves

In trees photosynthesis takes place in the leaves. The efficiency of leaves in the process and their hardiness in an extraordinarily com-

377 Influence of the environment on the structure of a Norway Spruce *(Picea abies)*: (left) — a trunk growing in a closed, shaded forest, (right) — two trees growing from the outset in the open.

378 Dissected sculpture of the crown of a Giant Fir *(Abies grandis)*.

379 Pagoda-like crown in a tropical forest.

378

379

380 Group of firs *(Abies alba)*; a crown resembling a stork's nest is a mark of very old trees.

Death of a fir *(Abies alba).*

Slashed trunk of the African tree *Harungana madagascariensis* with typical orange exudation.

Cross-section of the trunk of an oak *(Quercus)* with clearly evident annual rings, heartwood, sapwood and bark.

Columnar crown of a mountain species of spruce *(Picea schrenkiana)*.

Shape of young beeches *(Fagus sylvatica)* in a closed forest.

381

plex environment is perhaps due to their diversity of shape and physiological specialization. During the course of millions of years trees evolved leaves of two basically different types. Broadleaved trees have leaves with broad, flat blades on a flexible stalk while conifers are furnished with hard and narrow needles (Fig. 386 and 387). A further difference is the length of time they remain on the trees. In temperate and many tropical leaf-trees one generation of leaves remains on the tree for less than a year. Evergreen trees may retain individual leaves for as long as two to five years and thus they are never entirely bare. The longevity of conifer needles is even greater—those of the Norway spruce remain on the tree for six to nine years and it is therefore continually green. Only the Eurasian and American larches shed their leaves before the onset of winter and put out new ones every year.

382

381 Umbrella-like crown filled with numerous epiphytes.

382 Spherical crown of a tropical tree in Africa.

383

384 Buds of the Common Beech *(Fagus sylvatica)*.

383 Crown of the light-loving pine *Pinus pallasiana*.

385 Buds of the Sycamore *(Acer pseudoplatanus)*.

It would seem, then, that deciduous trees are better suited for regions with regular periods of cold or drought, bare branches apparently being better able to survive the effects of drought and frost and wind. One drawback, however, is the large consumption of raw materials required for the repeated development of leaves, and the fact that this development takes up part of the favourable growth season—which may be very brief in the north or in high mountains. During the course of evolution most conifers in the northern hemisphere developed evergreen leaves which natural selection perfected by providing structurally and physiologically resistant tissues. In evergreen conifers photosynthetic activity can continue at fairly low temperatures. In contrast, larches have evolved a different pattern, possibly because their ancestors grew in the extremely severe climate of cold continental Asia or some older pre-continent.

Even in the savanna woodlands, where they have to survive periods of drought, broadleaved trees follow the same pattern of behaviour. Most shed their leaves, though there are evergreen forms, as well as the peculiar *Acacia albida,* which is leafless during the rainy season and in full foliage during the dry season.

Needles are all very much the same, except for minor details. The leaves of broadleaved trees, however, exhibit marked diversity. They must be adapted to a wide range of climates—from the far north with a brief summer season, through the Mediterranean region with its cold winter and dry summer, to the monsoon regions of the tropics with a dry climate, and the damp tropics which are well supplied with water throughout the year. They differ in size, shape, pubescence, division and structure of the blade, length of the leafstalk, and, of course, colour. Perhaps the most typical leaf shape is ovate, about ten centimetres long and five centimetres wide, with a stalk about two centimetres long (Fig. 388 and 389).

It is quite amazing that many trees of quite different origin finally developed this form of leaf. This is one of the best examples of the gradual development of similarities in trees living in the same environment.

There are a great many other forms and sizes besides the typical pattern described

384

385

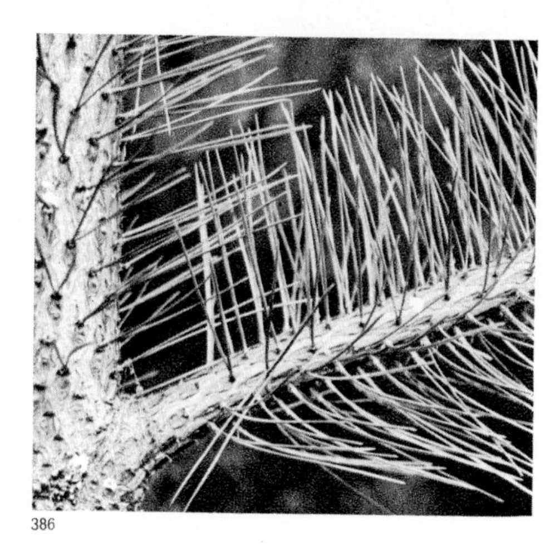

386

386 Characteristic arrangement of needles on the twig of the Scots Pine *(Pinus sylvestris).*

387 Characteristic arrangement of leaves on the twig of the Silver Birch *(Betula pendula).*

above. The lanceolate blade of the African *Anthocleista nobilis* is more that two metres long, whereas the leaves of Whistling Pine *(Casuarina)* are only small scale-like organs. The largest leaves of all are those of monocotyledonous trees—namely palm trees. These, as in *Raphia hookeri,* often measure more than ten metres in length. This is made possible by the superbly built stalk and rachis. Thus palms are able to do without a branched crown.

Light, drought and winds cause leaves to move in various ways that are important to the life of the plant and which would be hindered by a large leaf blade. For this reason the leaves of many trees are divided into separate leaflets. Leaves are veritable wonders of plant mechanics. They perform complex movements emanating sometimes from the base of the main leafstalk, sometimes from the base of the stalklets or even from the primary vein of the leaf blade. Thus the pinnate leaves of leguminous trees and shrubs, which are dominant in damp and dry tropical forests, are able to fold at noon to survive the heat and drought of midday with as little loss of water as possible. The giant *Piptadeniastrum africanum* of the

387

African rainforest folds its leaves at noon so perfectly that sunlight falls to the ground as in a clearing.

The surface of leaves may be smooth or leathery or they may be covered with glands or hairs (Fig. 390), the importance of which is the subject of debate. Thick hairs protect the leaf surface from being damaged by high temperatures and prevent excessive loss of water. They may also prevent damage by insects. Glands, on the other hand, attract small insects with their sweet secretions thus promoting the activity of animals that may help to pollinate the flowers, or preventing the spread of other species harmful to the leaves.

The leaves of forest trees are coloured various shades of green. Despite the presence of yellow and red pigments, the green pigment chlorophyll is the most important because it is involved in the photosynthetic process which transforms radiant energy into organic substances. The mixture of leaf pigments, however, sometimes gives the leaves a yellow or red tinge which makes such trees conspicuous. Everyone who visits a tropical forest for the first time is deceived by the young red leaves of broadleaved trees at the beginning of growth when they look more like flowers

388 Characteristic ovate leaf of the Goat Willow *(Salix caprea);* under surface on the left, upper surface on the right.

389 Difference in the shape of the leaf of two related species: left—Grey Alder *(Alnus incana),* right—Common Alder *(Alnus glutinosa).*

390 Pubescence on the underside of the leaf of a Downy Oak *(Quercus pubescens)* and gall of a gallfly of the family Cynipidae.

391

392

391 Flowering twig of the Common Beech *(Fagus sylvatica).*

392 Male catkins of Durmast Oak *(Quercus petraea).*

393 Panicle of male flowers of the Common Ash *(Fraxinus excelsior)* in early stage of growth.

394 Hornbeam *(Carpinus betulus).*

than leaves. The same is true of the whitish foliage of some cecropias.

Seed production

The ability of forest trees to survive depends on their ability to bear flowers and produce viable seeds. The shape of flowers, their movements and their associations with pollinating agents are supreme examples of nature's ingenuity.

From man's viewpoint European and North American species have very insignificant flowers as so many of them are pollinated not by insects, but by the agency of the wind. Such flowers are generally characterized by the absence of sepals and petals which are so conspicuous a part of insect-pollinated flowers, and frequently the flowers appear before the leaf buds unfold. The European beech (Fig. 391), for example, which produces flowers at the same time as the leaves, has two kinds—long, stalked, male catkins and two-flowered, upright female spikes. Beeches flower at intervals of five to ten years when they are more than sixty years old. Oaks bear pendant male catkins (Fig. 392) and female flowers are borne either singly or several to a stalk (which

393

394

grows from the axils of the current crop of leaves). Oaks, too, bear flowers only when they reach the age of about fifty, producing them at intervals of three to seven years. Hornbeams (Fig. 394) begin flowering at the age of thirty or forty and then flower practically every year. Ashes have flowers of different kinds (Fig. 393), which are borne in panicles and which may be bisexual, or only male or female. Some trees may produce only male flowers, others only female flowers. It is not known what influences the differentiation of the sexes.

Willows produce flowers annually and in great profusion from their earliest youth (Fig. 396). The catkins usually appear at the same time as the leaves and when the seeds are ripe they fall and cover the ground. This is a typical characteristic of light-loving pioneer trees and

395

395 The Common Horse Chestnut *(Aesculus hippocastanum)* has ornamental flowers and foliage.

396 Catkins of the Goat Willow *(Salix caprea)*.

396

shrubs which invade every spot where the canopy of the forest is broken, allowing light to enter.

Only a few European trees and shrubs have flowers that are ornamental or have a pleasant fragrance. The Common Horse Chestnut *(Aesculus hippocastanum)* bears rich, upright panicles of flowers arranged in whorls (Fig. 395). They are coloured white with red and yellow blotches and are pollinated by insects. When one takes into account the decorative, palmately compound leaves and glossy brown seeds which are nourishing food for wild game then it is not surprising that this forest tree is popularly cultivated even far beyond the limits of its original range of distribution. Handsome trees are also found in North America. One such is the tulip tree which has large green flowers with orange spots.

The reproductive organs of conifers are not flowers or inflorescences in the true sense of the word and thus it cannot be held against them that they are sometimes rather inconspicuous. They are produced only by older trees at intervals of several years, and consist of separate male and female cones (Fig. 398). In a closed forest Norway spruce bears cones only after reaching the age of sixty and at intervals of four to five years. The pollen grains of spruces, firs and pines are furnished with air sacs that facilitate dispersal by the wind, which is the chief pollinator of conifers.

In tropical forests the predominance of wind-pollinated trees is not as marked as in temperate forests. Important pollinators in the tropics are insects (butterflies and moths, beetles, bees and flies), birds and bats, which is why the flowers are more segmented, more colourful and more fragrant than in Europe and North America. Furthermore, the position of the flowers on the trees is quite unusual compared with that of flowers on wind-pollinated trees. In many species they grow directly on the trunk (cauliflory) (Fig. 397), which enables insects flying at trunk level to come into contact with the flowers and pollinate them. Other tropical trees bear flower clusters on stalks that hang from the base of the trunk to the ground, thus enabling pollination by insects crawling in the forest litter (Fig. 401). Species pollinated by bats often have flowers on long stalks suspended from the branches and hanging freely in space to facilitate landing of the pollinators. Best known of these species is the Sausage Tree (so called because the fruits resemble sausage links)—*(Kigelia*

397

397 Cauliflory: flowers on the trunk of the tropical *Omphalocarpum ahia.*

398 Young female cone of Norway Spruce *(Picea abies).*

398

399

399 Ripening fruits on the trunk of *Omphalocarpum ahia;* Ghana.

400 Fruit of *Allexis cauliflora* (Violaceae) from which the seeds shoot out to all sides.

401 *Chytranthus villiger* of the soapberry (Sapindaceae) family produces flowers at the base of the trunk.

africana) of the Bignoniaceae family. At night the flowers of tropical trees may be visited successively by various species of fruit bats (Megachiroptera) and towards dawn by scavenging insects. Most New World hummingbirds (Trochilidae), on the other hand, as well as sunbirds (Nectariniidae) of the Old World tropics visit flowers during the daytime.

The fruits or seeds of forest trees and shrubs also exhibit marked diversity. The forms of seeds that have been developed by natural selection during the course of evolution provide efficient dispersal and a good start in life for the young seedling. Most seeds and fruits are adapted for dispersal by wind but in many instances they are equipped with special mechanisms for dispersal either under their own power or by animals.

In the case of tall trees in the dense tropical forests, most important is the weight of the fruit which must fall through the canopy of branches and foliage to the ground. The fruits of *Omphalocarpum ahia* weigh more than a kilogram (Fig. 399), so they fall through the branches with great speed and land on the ground with a thud. Because of this chara-

400

401

402

402 Ripening fruit of *Hura crepitans* of the spurge (Euphorbiaceae) family, which explodes when ripe, scattering the seeds far and wide; Cuba.

403 In mangrove woodlands the young seedlings of *Rhizophora racemosa* grow on the parent tree and when they fall become embedded deep in the mud; Tanzania.

403

cteristic the people of Ghana call this tree berebere-tim (falling quickly). The seeds of other trees are dispersed by forces developed by the fruits as they dry. The tree *Allexis cauliflora* of the Violaceae family, which produces flowers and fruits directly on the trunk, shoots seeds from the open capsule up to five metres from the base of the trunk through the undergrowth of the rainforest (Fig. 400). Even more impressive is the performance of *Hura crepitans* of the Euphorbiaceae family of the tropical forests of America. When fully ripe its woody fruits (Fig. 402) explode and the seeds are scattered like pieces of shrapnel. The pods of many leguminous trees burst and scatter their seeds in a similar way. In mangroves the seeds germinate while still attached to the parent and after a time the seedling, which looks like a pointed peg, falls pointed end first into the mud where it becomes permanently established (Fig. 403).

Many fruits are palatable to animals, and ensure their dispersal by this means. Some are carried by animals to distant hideaways, while others are able to pass through an animal's digestive tract undamaged and germinate in the fertile medium of its faeces far from the parent tree. Examples of such fruits are the banana-like fruits of the tropical *Cola chlamydantha* of the Sterculiaceae family (Fig. 405).

The fruits of most European broadleaved trees are equipped for gliding or for being carried by air currents. The ripe nutlet of the com-

404

405

404 Fruiting twig of the Hornbeam *(Carpinus betulus).*

405 Fruit of *Cola chlamydantha;* Ghana.

406

406 Large-leaved Lime *(Tilia platyphyllos)*: fruit with bract.

407 The small weight and special shape of the one-seeded samara of the Silver Birch *(Betula pendula)* contribute to its rapid dispersal by wind and water.

407

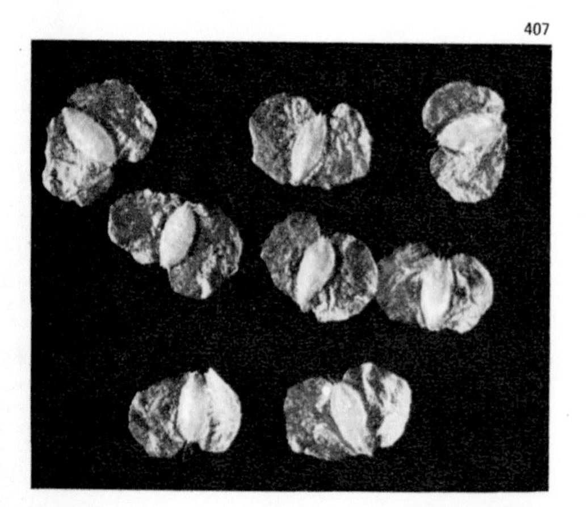

mon hornbeam, which is furnished with a three-lobed, wing-like bract, turns like a corkscrew as it is carried far beyond the edge of the tree (Fig. 404). The fruit of the lime tree likewise has a bract that acts as a wing for seed dispersal (Fig. 406). Such a winged seed is called a samara. Maples have winged double samaras that are easily carried by the wind. The small, light samara of the birch, furnished with two membranous wings, not only glides through the air but also floats equally well on water (Fig. 407), which explains why birches are the vanguard of the forest even in places many kilometres distant from the seed-bearing trees. The related alder (Fig. 408) has very small, flat samaras capable of being carried great distances by air or water.

Conifers spread either by means of individual seeds or by entire cones. The berry-like cones of junipers are a favourite food of birds which disperse them rapidly in pastures and spaces cleared by the felling of a forest (Fig. 409). Birds are also important in spreading yews, the seeds of which have a bright red fleshy covering. The dominant coniferous species of North American and European forests spread by means of seeds which are furnished with a wing that enables them to slip and glide through the air. The Weymouth pine disperses its seeds far and wide (Fig. 410) and its offspring inhabit a wide range of habitats. From the early age of twenty, this pine produces seeds every three or four years. The seeds remain viable for up to three years and a very high percentage germinate. Furthermore, seed-producing trees may live to be up to 500 years old, so it is not surprising that this species has spread with great vigour.

The strategy employed by larches (Fig. 411) involves the cones opening in succession during the first and sometimes during the second year after they ripen, thus increasing the possibility of the seeds falling during a period favourable for germination.

One of the most successful trees is the Scots pine. It reaches maturity at the age of thirty (in open country as early as 15), and produces seeds at intervals of three to four years. The female cones (Fig. 414) mature slowly on the tree so that the winged seeds are not released until the spring of the third year. Atmospheric moisture determines when the ripe cones open. In damp conditions they remain closed, but when they become dry, they open and release the seeds which float

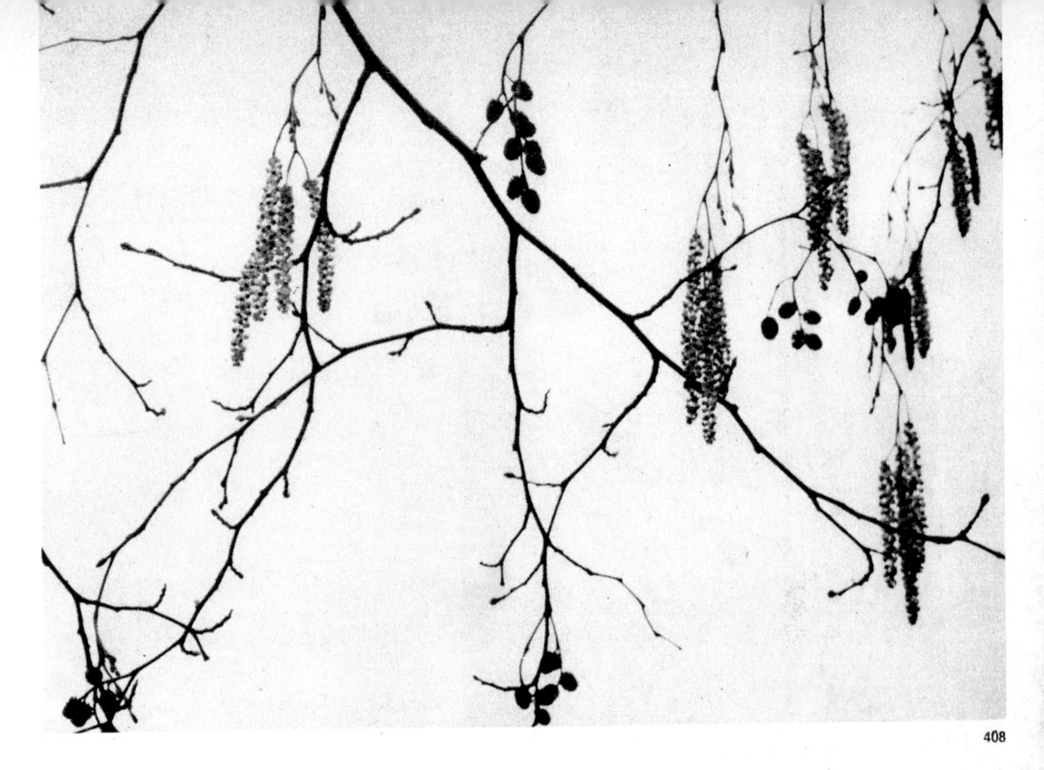

408 Flowering catkins and previous year's cones on the branches of the Common Alder *(Alnus glutinosa)*.

through the air. The crown of the Scots pine is constantly full of cones in various stages of ripeness and the ground beneath is covered with thousands of empty cones which have dispersed their seeds far and wide. The rest depends on the right combination of light, moisture and food for the seeds to germinate, survive the first critical weeks as seedlings, and become established in their new home. The cones of the spruce also open and close in response to dry or damp conditions (Fig. 412 and 413).

Trees that tolerate shade in youth can grow under the protection of the parent tree (Fig. 415 and 417), but as a rule they are less vigorous and have a slow rate of growth. Their opportunity comes when the aged parent dies, leaving an empty space in the closed canopy of the forest. In the shade of the virgin tropical forests there are thousands of such seedlings awaiting their turn (Fig. 416) — inconspicuous and delicate plants with genes in which is coded the possibility of becoming a giant of the forest ecosystem.

409 Berry-like cones of the Common Juniper *(Juniperus communis)*.

410 Cone of Weymouth Pine *(Pinus strobus).*

411 Ripening cones of the Common Larch *(Larix decidua).*

410

411

412 Empty cones at the foot of a spruce—the seeds are scattered in the vicinity.

413 Spruce cones open in dry weather.

414 Cones of Scots Pine *(Pinus sylvestris)*.

412

413

414

415

415 Young spruce growing in the shelter of a parent branch.

416 Seedlings in a tropical virgin forest may grow into veritable giants.

417 New growth of spruce in a gap in a mountain forest.

416

417

418

418 The Common White Hellebore *(Veratrum album)* is the most robust of the liliaceous herbs of the temperate forests of Europe.

THE OTHER PLANTS OF THE FOREST

The plants of all ecosystems in the world include both giants and dwarfs—species that are dominant in size and number and species that are small and inconspicuous. The importance of the members of an ecosystem is not determined by the weight, size, or number of individuals but by their role in the flow of energy and the cycling of nutrients. Quite important may be the role of plants that are inconspicuous or even so small that they are visible only with the aid of a microscope.

In a forest, trees are the giants of the plant kingdom, but numerous studies have confirmed that viruses, bacteria, algae, blue-green algae, fungi, lichens, liverworts, mosses, club mosses, horsetails, ferns, grasses, subshrubs, shrubs, vascular parasitic and hemiparasitic plants, stranglers, epiphytes and other life forms have their own special niche in the forest community.

A great many forest plants are closely bound to the forest environment and occur elsewhere only occasionally and under special circumstances. Stands of forest trees take up a large amount of space both above and below the ground. Their crowns, trunks and roots divide this space into many layers and separate habitats where natural selection has resulted in the evolution of ever newer species of cryptogams and phanerogams (non-flowering and flowering plants). The presence of these new types further augments the diversity of the forest environment in time as well as in space.

The microflora

All the surfaces, crevices and cavities in the forest are occupied by microscopic bacteria. Whole populations of specialized bacteria, such as *Bacterium herbicola-aureum* and *Pseudomonas fluorescens,* live and multiply on the surface of green leaves and stems. Apart from certain soil microorganisms, these species are perhaps the most striking cosmopolitans. The same species may be found on foliage in all types of forest in all parts of the world.

Soil microflora, which includes various bacteria, fungi, green algae and blue-green algae, is greatly variegated and includes many more species (Fig. 419). The greatest diversity and the greatest numbers are to be found in the humus-rich upper layers which have a favourable soil structure and an abundance of nutrients and water thus providing sources of energy for organisms with different requirements.

Each gram of soil contains millions of bacteria. Soils rich in nutrients, moisture and oxygen may contain several milliards of them. Such numbers are estimated both by counting bacterial cells under a microscope and by counting growing colonies on agar plates.

Soil bacteria are classified in many ways. For example, they may be divided into groups on the basis of the way in which they obtain energy. Autotrophic species obtain energy from mineral substances, and heterotrophic species require the energy-rich substances of other organisms for their existence. They are also classified on the basis of whether they are able to live where there is no free oxygen (anaerobic species) or whether they are able to live only where free oxygen is present (aerobic species). They are further grouped according to their method of propagation into those that besides multiplying by vegetative means also produce spores and those that multiply only by vegetative means. Another important aspect is whether the bacteria obtain nourishment from living organisms (parasitic species) or live on organic material from dead plants or dead animals (saprophytic species). Bacteria are divided on the basis of the material they decompose, into those that decompose the most voluminous substances of the forest, namely cellulose, and those that decompose proteins. On the basis of how they process nitrogen, the most important nutrient in the entire forest system, they are classified as, for example, nitrifying, denitrifying and nitrogen-fixing bacteria.

Soil bacteria ensure the smooth and continuous metabolism of the forest ecosystem. Without their activities as decomposers life in the forest would soon come to a halt because most of the nutrients would have been exhausted from the soil and would be concentrated in the dead plant and animal matter on the surface. Forest trees and herbs are provided with mineral nutrients as well numerous vitamins, growth substances, and antibiotics. Furthermore, certain bacteria are able to fix atmos-

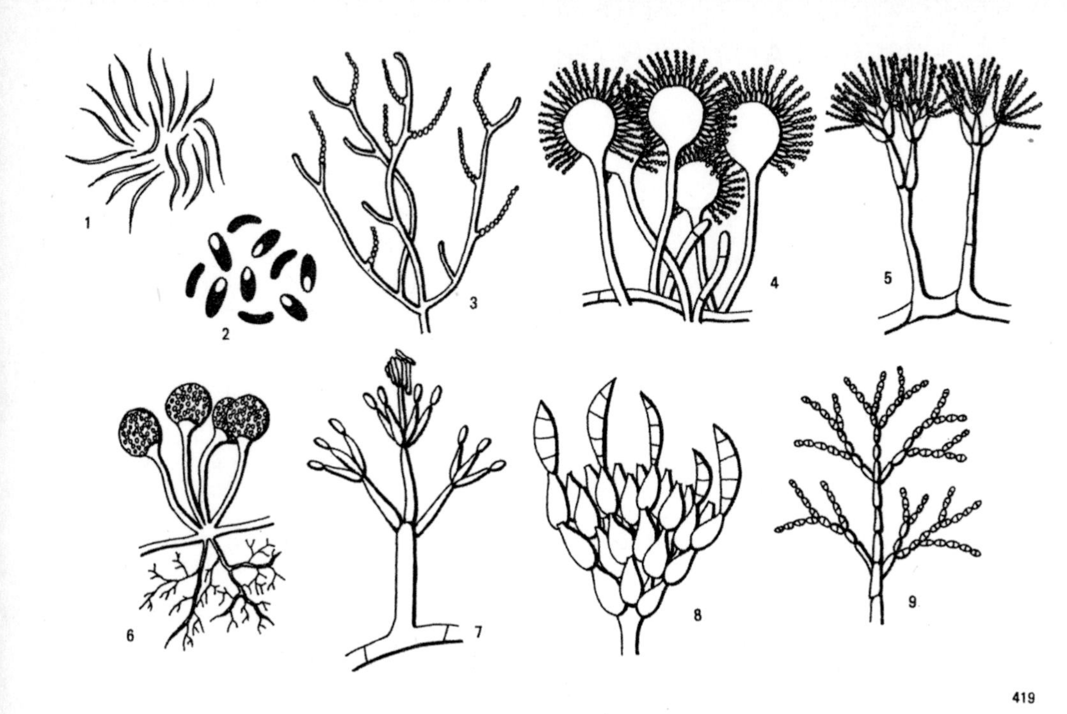

pheric nitrogen. These include species living freely in the soil, such as members of the genus *Azotobacter, Clostridium,* and the tropical genus *Beijerenckia,* and bacteria that live in a symbiotic relationship on the roots of leguminous plants *(Rhizobium).* The activity of nitrogen-fixing bacteria provides the forest with new continual supplies of nitrogenous fertilizer.

Many dominant trees of tropical forests, namely members of the families Papilionaceae, Caesalpiniaceae, and Mimosaceae, contain symbiotic bacteria that make atmospheric nitrogen available for their use. Bacteria of the genus *Rhizobium* make small nodules on the roots of trees from which nitrogenous substances pass directly into the tree sap as well as into the surrounding soil, indirectly enriching the forest ecosystem with nitrogen when the roots and nodules die and decompose. Certain symbiotic fungi (Fig. 420) are of similar importance to forest trees the same as blue-green algae that live in the roots of tropical and subtropical cycads.

Algae and blue-green algae live on damp leaves, branches and trunks as well as in the upper soil layers of the forest (Fig. 422). Some only grow on trees, conferring neither

419 Top row—bacteria of forest soil. Middle and bottom rows—moulds and mildews of forest soil, showing spore-producing organs:
1—bacteria that break down cellulose,
2—bacteria that fix atmospheric nitrogen,
3—Actinomycete, 4—*Aspergillus,*
5—*Penicillium,* 6—*Rhizopus,* 7—*Verticillium,*
8—*Fusarium,* 9—*Cladosporium.*

420 Tangle of roots of the Common Alder *(Alnus glutinosa)* with nodules inhabited by *Plasmodiophora alni* which fixes atmospheric nitrogen.

harm nor benefit, but others are parasitic and these weaken the host. In practically all damp forests the moist bark of the trees is coated with the green alga *Pleurococcus vulgaris*. In damp tropical forests, algae of the genus *Trentepohlia* and various blue-green algae *(Hapalosiphon, Phormidium, Scytonema)* are commonly found on the blades of green leaves, where, together with fungi, lichens and mosses, they form a complex leaf community.

As well as blue-green algae, green algae and diatoms are commonly found both on the surface and within the soil. In damp weather they form a spreading green carpet on the surface. The number of algae rapidly decreases with depth in the soil but they may still be encountered as deep as two metres where they have been carried by percolating water. In the dark underground depths algae and blue-green algae cannot make their own food by photosynthesis for no light penetrates there. Experiments have proved that in such a case soil algae are able to carry out nutrition heterotrophically and utilize, for instance, the sugars produced by the roots of higher plants.

The number of individual algae in the soil may be lower than the number of bacterial cells, but even so they run into hundreds of thousands per gram of soil. The importance of algae populations to the fertility of forest soil is truly great: the dying bodies of algae and blue-green algae form a significant amount of humus; their mucilage cements fine soil particles and helps to bind the soil; during the process of photosynthesis they release into the soil oxygen necessary for the respiration of roots and all soil organisms; and they excrete into the soil various organic substances that promote the growth of the roots of higher plants, soil fungi and soil bacteria. Soil algae, for instance, promote the activity of nitrogen-fixing bacteria in the soil and blue-green algae themselves include species that are able to fix atmospheric nitrogen within the soil and thus contribute directly or indirectly to the nutrient cycle of the forest ecosystem.

420

421

421 The bracket fungus *Fomitopsis pinicola* on a Norway spruce with exuded drops of colourless fluid.

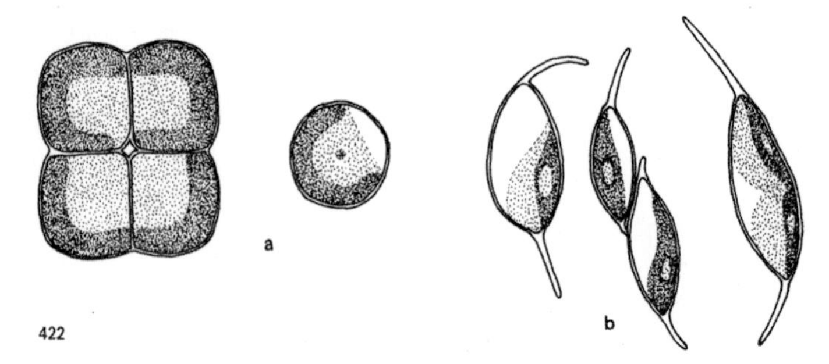

422

422 Algae that coat plant surfaces: a — *Pleurococcus vulgaris,* which lives on the bark of trees, b — *Keratococcus bicaudatus* which lives in the surface layer of the fruiting bodies of bracket fungi.

Fungi

Forest fungi are certainly more numerous and diverse than the ordinary mushroom picker, who collects broad-capped edible basidiomycetes such as Boletus and Lepiota, might think. It is probable that in most types of forest there are up to a hundred times more fungi than vascular plant species, though on a world scale vascular plants far outnumber fungi.

Important in all layers of the forest are two types of fungi, perhaps best called the moulds and mildews, which live as saprophytes on dead organic matter or as parasites in the cells and tissues of plants and animals. A typical parasite is the mildew which coats the leaves of trees with a white mycelium which develops haustoria that penetrate the leaf tissues and syphon off the nutrients it requires. Included in their number are various species that live high up in the crowns, as well as in the soil.

Soil-dwelling moulds play an important role in the decomposition of organic matter and at the same time produce important substances which may have a stimulating or toxic effect on bacteria and other fungi. Since World War II this characteristic of fungi has been exploited by the pharmaceutical industry in the production of antibiotics, the best-known of which is penicillin. In the forest ecosystem, however, these substances have been influencing the environment for millions of years and are one of the agents that regulate the stability of this ecosystem.

The fungal inhabitants also include the yeasts of the genus *Saccharomyces* that live in the sweetish liquids of forest plants, round floral and extrafloral nectaries, in fruits and

The Fly Agaric *(Amanita muscaria)* in drying undergrowth of an autumn forest.

Veils of the moss *Orthotrichidium perpinnatum* in a mountain mist forest.

423 Pendant fruiting bodies of a basidiomycete of the species *Cyphella digitalis* on the branch of a fir in a virgin forest.

424 The bracket fungus *Polyporus umbellatus* is a favourite food.

423

424

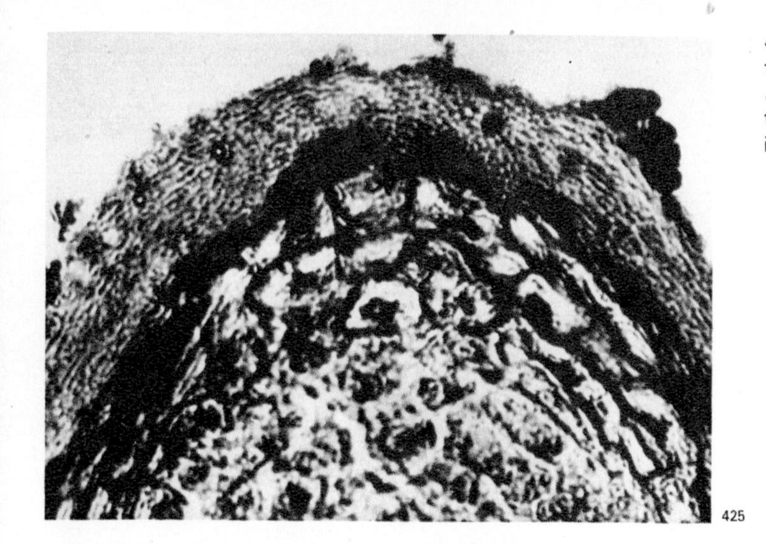

425

425 Ectomycorrhiza on the African tree *Afzelia africana;* clearly visible on the surface of the root tip is the mass of mycelium.

426

426 Members of the large genus of edible boletus *(Boletus edulis)* are the most widely gathered mushrooms in Europe's forests.

427 The Stinkhorn *(Phallus impudicus)* in the litter of an oak forest.

428 The parasitic *Boletus parasiticus* on the fruiting bodies of *Scleroderma vulgare*.

427

428

429

429 *Dictyophora phalloidea* grows in tropical forests.

430 The branches of trees in mountains are often covered with *Usnea* lichen; mist forest on Kilimanjaro, Africa.

430

431

431 The coloration and structure of the bark of tropical trees is changed by crustaceous lichens; tropical Africa.

432 *Polytrichum formosum* is common in shady forests and on rooting trees.

wherever the sweet sap that flows in the phloem of higher plants finds its way to the surface. Yeasts obtain the energy they need by fermenting sugars to form alcohol. The influence of these microscopic yeasts on the structure of the forest may have very direct effects. When the fruits of the tree *Detarium microcarpum* of Africa's savanna woodlands begin to ferment, elephants eat them en masse, with the result that they often become intoxicated. Under the influence of alcohol they break off crowns and uproot trees in the surrounding forest. In some areas the action of yeasts is put to good use by man. The inhabitants of tropical forests collect palm sap high up beneath the crown of leaves and within a single day they have fermented palm wine which can be consumed immediately.

Of quite different appearance, close relatives of the mildews and yeasts include the morels *(Morchella)* and the false morels *(Helvella)* in which the fruiting body looks rather like a toadstool, as well as such species as the Orange Peel fungus *(Peziza)*. Also included in this group are truffles *(Tuber* sp.*)* which form mycorrhizae with the root cells of trees and have underground, potato-shaped fruiting bodies which are especially prized on the Continent.

The soil as well as the dead and living tissues of trees in the forest are laced with the mycelia of basidiomycetes—the toadstools and their relatives. These include the well-known bracket fungi *(Polyporus* sp.*)* which have fruiting bodies growing like shelves from the trunks of old trees and stumps in virgin forests (Fig. 421 and 423). Most are parasitic fungi that harm the host trees, but many are partly or entirely saprophytic. The fruiting bodies of bracket fungi are a decorative feature of the forest and some are gathered as food, while others have been used as medicinal drugs. For example the bracket fungus Surgeon's Agaric *(Fomes fomentarius)*, common on beeches and oaks in European and North American forests, yields a pulp that was used in former times in surgery, to halt bleeding. Another medicinal species is *Polyporus officinalis*, common on larches in the Siberian taiga and subalpine forests of the Alps. A popular

432

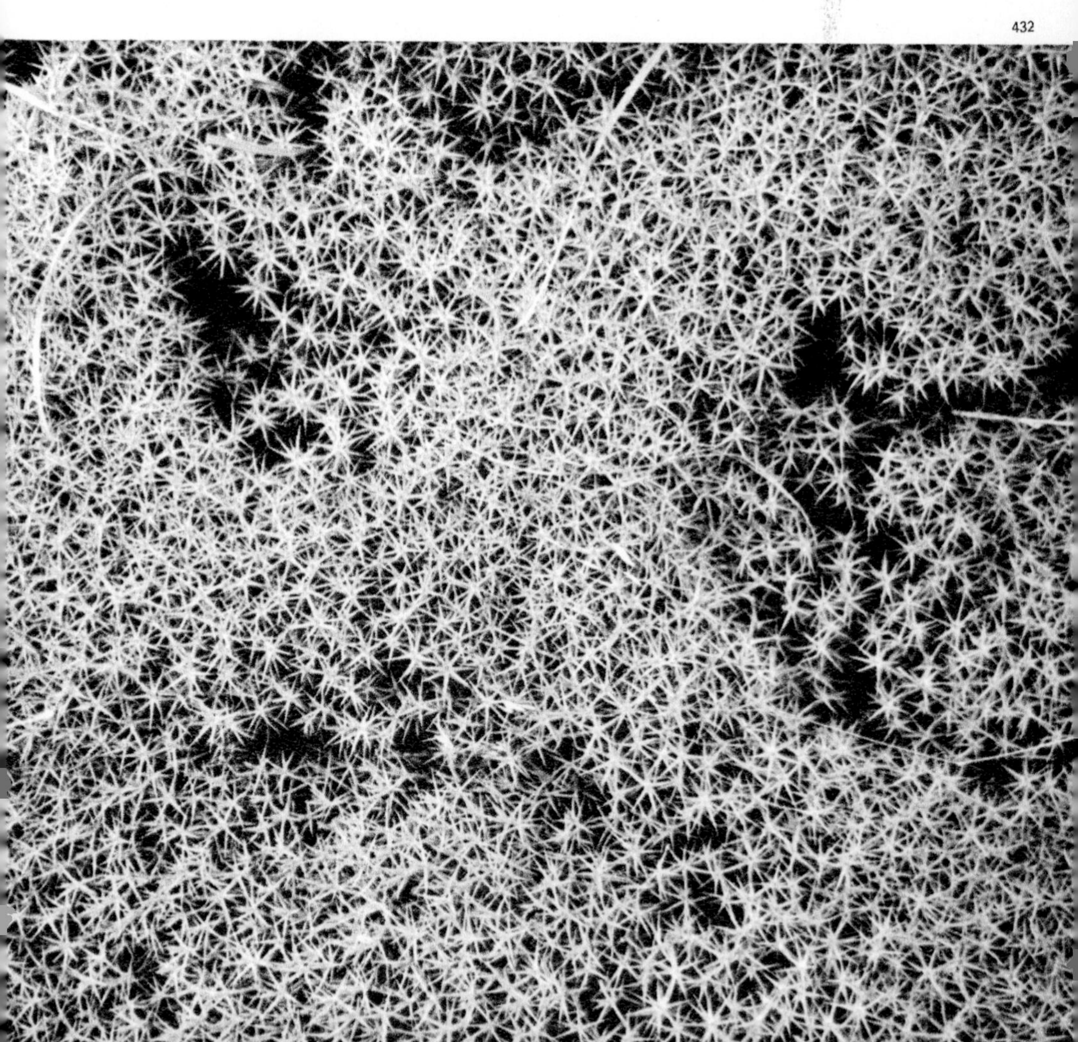

433

edible species is *Polyporus umbellatus,* which has a pleasant smell and also a fine taste when cooked (Fig. 424).

Toadstools are mostly saprophytic or symbiotic organisms (Fig. 426). Their fruiting bodies emerge from various substrates of the forest ecosystem and are soon inhabited by a large community of plants and animals that obtain nourishment from them. In temperate forests the fruiting bodies emerge only during the growth period, particularly in late summer and autumn. Their variety of colour and shape are among the most interesting aspects of the forest ecosystem. In the tropics the ap-

433 Cushions of the moss *Leucobryum juniperoideum* spread over the acid humus of a mountain forest.

434 Carpet of the club moss *Lycopodium annotinum* in a coniferous forest.

434

435

436

435 The Hard Fern *(Blechnum spicant)* grows in the coniferous forests of European mountains.

436 A ray of sunlight falls on the fern *Dryopteris filix-mas.*

437 Diagram showing vegetative reproduction of the tree fern *Cyathea manniana* which grows in the tropical forests of Africa: A—trunk with crown of leaves, cone of surface roots and underground roots, B—vegetative reproduction by means of lateral branches which root in the soil.

pearance of the fruiting bodies of toadstools is not synchronized and those of a particular species appear singly at any time during the course of the year. It would seem that mushroom picking, a popular pastime in European forests, would hardly be worth the effort in the tropics.

As was explained in the preceding chapter, many toadstools form mycorrhizae together with the terminal roots of trees. The mycelium on the surface of ectomycorrhizae resembles cortical tissue (Fig. 425) but differs markedly in size of cell and physiological activity. The morphology and anatomy of mycorrhizae depend firstly on the structure of the root and secondly on the fungal component—the species of fungus that participates in the symbiotic association. Some basidiomycetes are restricted to a certain species of woody plant (or limited range of species)—for instance *Leccinum scabrum* is associated only with birch. Other species such as *Boletus, Amanita,* etc., can form mycorrhizae with a number of woody plants. Some woody plants are similarly restricted and are able to form mycorrhizae only with a single species of fungus, though most form such symbiotic associations with a number of fungal genera.

The complex relationships in the forest ecosystem are illustrated also by the parasitic *Boletus parasiticus* which penetrates the mycelium into the fruiting body of the Common Earth-ball *(Scleroderma vulgare)* and there forms its own fruiting body (Fig. 428).

The fruiting bodies of forest fungi exhibit great diversity and are striking not only in shape and colour but sometimes also in their distinctive smell. One species noted for its repellent smell is the Stinkhorn *(Phallus impudicus),* found in the forests of the temperate zone. Noted for its beauty is *Dictyophora phalloidea,* which grows in tropical forests (Fig. 427 and 429).

Lichens

Lichens are plants that grow on the surface of the soil, on the bark and branches of trees, and occasionally also on green leaves. They are dual organisms consisting of a fungus mycelium which encompasses algal cells. Lichens are crust-like, leafy or upright-growing in form and affect the life of the forest only

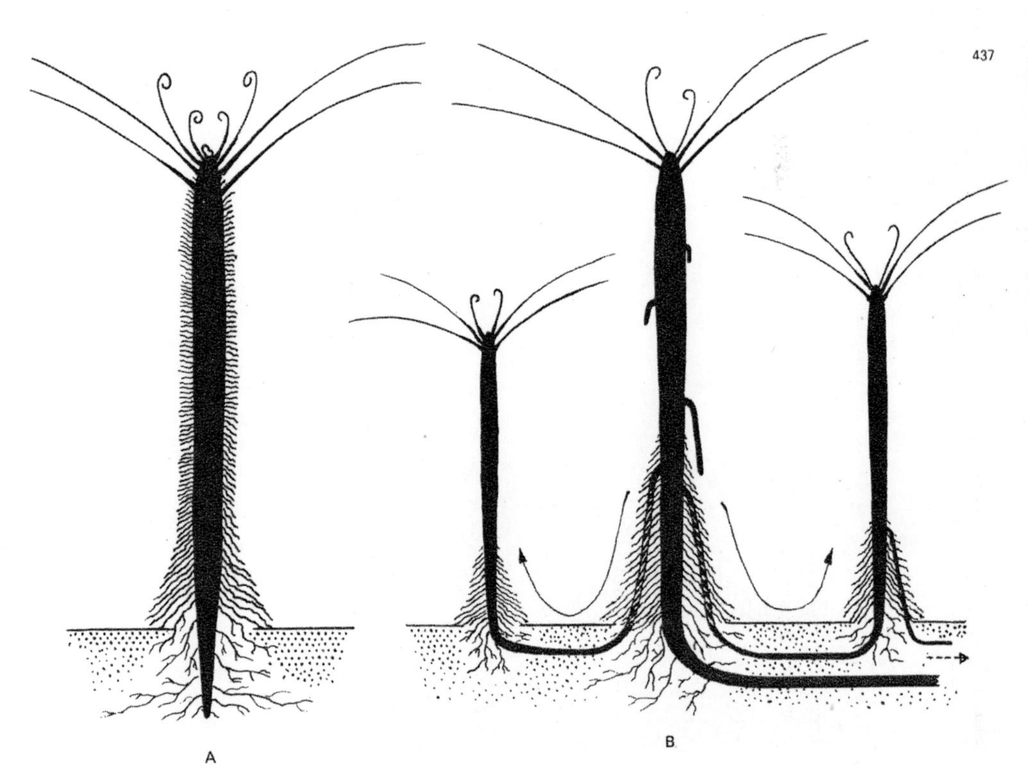

438

439

438 Lianas and mosses have made it easier for the Spleenwort *(Asplenium africanum)* to gain a foothold on the trunk of a tropical tree.

439 The tree fern *Cyathea manniana* in a shady ravine of a tropical forest.

440 Branch of a tropical tree with the epiphytic fern *Drynaria laurentii* and orchids of the genus *Bulbophyllum*.

440

indirectly though they have a marked effect on its appearance. The crowns of trees in mountain forests are often covered with *Usnea, Evernia, Parmelia, Xanthoria* and many other brightly-coloured lichens (Fig. 430 and 431). It is the lichens that give virgin forests throughout the world their patina and create the impression of an old and stabilized ecosystem. In industrial regions this forest patina rapidly vanishes because lichens do not tolerate atmospheric pollution, particularly sulphur dioxide. Even in the mountains of central Europe lichens are becoming extremely rare even although industrial centres are hundreds of kilometres distant. Lichen desert is the name given by ecologists to places where lichens are no longer present. A desert containing forest trees may seem a somewhat exaggerated term but from the viewpoint of environmental evolution it represents a serious warning.

Bryophytes

Faithful companions of forests throughout the world are the bryophytes (mosses and liverworts). In damp situations they form green coats, cushions and carpets on the ground, on stones, on the branches of trees and on small

441

441 The Wood Anemones *(Anemone nemorosa)* brighten broadleaf forests in spring.

442 The strangler *Ficus leprieuri* beginning to spread over a huge tree in the tropical forest.

443 The Chickweed Wintergreen *(Trientalis europaea)* grows in the peat and mountain coniferous forests of Europe.

442

443

crags. They win despite the strong competition of herbs and grasses, mainly in places which are very moist and shaded, where there is little fine earth and where there is an accumulation of acidic humus. However, not all liverworts and mosses have such minimal requirements as to inhabit those spots where trees and herbs are unable to exist. Some bryophytes are very demanding and their presence is an indication that the site is a fertile one.

Liverworts (Hepaticae) and mosses (Musci) are plants composed of complex tissues which show only the beginnings of differentiation into root, stem and leaves. They include types which are flat and leafy and pressed to the substrate and types which possess stem and leaf-like structures. Mosses always have leafy stems attached to the substrate by filaments called rhizoids, which also absorb water solutions.

The most congenial conditions for the

444 Three stages in the development of a strangler of the genus *Ficus* growing in the crown of a tropical tree.

development of bryophytes are provided by the mountain mist forests of the tropical zone, by the waterlogged and nutrient-deficient forests of the tropical and temperate regions (various kinds of peat forests) and by the coniferous taigas of North America and Eurasia. Comparatively few bryophytes are to be found among the ground flora of true tropical rainforests and they are entirely absent in the dry savanna woodlands as well as the sclerophyllous forests of the Mediterranean region. In mist forests the crowns are thickly covered with bryophytes and some, such as *Ortostichidium perpinnatum* in the mountains of Africa, hang like ribbons in thick sheets. On the old branches of giant trees, mosses form a carpet for small suspended gardens in which epiphytic orchids, bromeliads and ferns grow. Such carpets are formed, for example, by *Macromitrium levatum* in the equatorial forests.

Bryophytes cover the surface of the soil in the vast peat forests of Malaysia and the Amazon basin. In Europe and North America the dominant species in peat forests are chiefly of the genera *Sphagnum* and *Polytrichum* (Fig. 432). The raw and acidic humus of coni-

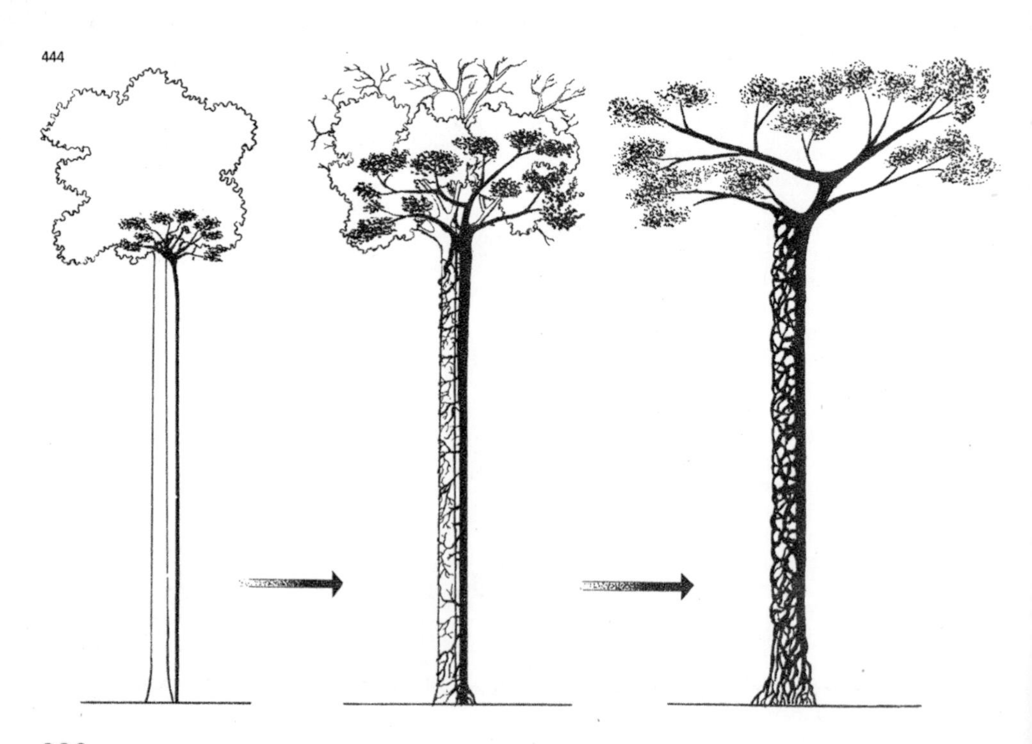

Rosebay Willowherb *(Epilobium angustifolium)* — dominant plant of clearings in the cold regions of North America and Europe.

The climbing shrub *Mussaenda erythrophylla,* called Ashanti Blood by the people of Ghana.

The Dogtooth Violet *(Erythronium americanum)* in the undergrowth of a mixed forest.

Spring flower — *Trillium erectum.*

Thousands of Snowflakes *(Leucojum vernum)* grow in forests.

445

445 The Common Hop *(Humulus lupulus)* and a carpet of Stinging Nettle *(Urtica dioica)* in a flood plain forest on the banks of a river.

446

446 Detail of the leaves and cones of the Common Hop *(Humulus lupulus).*

447 The One-flowered Wintergreen *(Pyrola uniflora)* is an evergreen plant of moist coniferous forests. It has a characteristic endomycorrhiza.

447

ferous forests is favourable to the growth of many leafy liverworts and mosses of the genera *Bryum, Dicranum, Hylocomium, Leucobryum* (Fig. 433). *Plagiothecium* and *Polytrichum.*

An abundance of mosses and liverworts is very important to the cycling of water and nutrients in the forest. The moss cover protects the soil from drying out which is why thriving seedlings are often to be found in a carpet of bryophytes. Bog mosses *(Sphagnum)* are particularly well adapted for holding water because their stems and leaves contain specialized cells which have great powers of absorption and retention. Sometimes, however, an excessively thick carpet of mosses hampers the germination of seedlings and regeneration of the forest. Compact cushions of *Leucobryum glaucum* in poorly cultivated spruce or pine forests are usually associated with the unhealthy development and progressive degradation of the forest.

Ferns

A large number of members of the fern family (Phylum Pteridophyta), which includes the club mosses (Lycopodinae), horsetails (Equisetinae), and ferns (Polypodiinae), are forest dwellers. Millions of years ago some of their ancestors were tall trees which formed forests particularly in swampy and permanently waterlogged places. Present-day pteridophytes are comparatively insignificant relics of a group which was conspicuous and dominant in the past. In spite of its name and appearance the club moss *(Lycopodium annotinum)* is not a true moss. Its club-like spore case (sporangium) is comparable with the sporangia borne by ferns (Fig. 434). Similar structures are found on horsetails, and some species such as *Equisetum sylvaticum* still live in swampy areas. Present-day ferns are of comparatively modest size, most of those occurring in European and North American forests averaging less than a metre in height (Fig. 435 and 436). In shady forests only the Royal Fern *(Osmunda regalis)* and Bracken *(Pteridium aquilinum)* reach a height of two metres.

Found in tropical forests and very occasionally also in southern Chile, South Africa and New Zealand are the tree ferns of the Cyatheaceae family (Fig. 439). They are only distantly related to the giant trees of the Carboniferous forests which formed the coal deposits that are used today. Existing tree ferns have an upright, unbranched stem, usually slender (up to twenty centimetres thick) and less than ten metres tall. (*Cyathea kermadecensis,* however, is up to twenty-five metres tall and fifty centimetres thick.) The stem is terminated by a crown of pinnate leaves often several metres long. In the shaded ravines of tropical mountain chains ferns of this family are very common and together with palm trees are characteristic features of tropical forests. They number some 800 species of which about 300 belong to the genus *Cyathea* and a similar number to the genus *Alsophila.* The African species *Cyathea manniana* is noted for its vegetative means of reproduction. The stem

448 Flowering rhododendron at the upper timberline in the Himalayas.

449

449 Pine forest on peat soil, and undergrowth of the Wild Rosemary *(Ledum palustre).*

450 The White Butterbur *(Petasites albus)* on the banks of a stream in a mountain beech forest.

451 Fruits of the Bilberry *(Vaccinium myrtillus).*

450

451

produces a branch, even at a height of several metres, which arches to the ground where it takes root and forms a new upright stem (Fig. 437).

Tropical forests are the home of many other large ferns, such as vigorous species of *Athyrium, Cyclosurus, Gleichenia, Marattia, Pteris* and *Dryopteris* and many smaller, more delicate ferns that are not content only with the shade on the forest floor but may also be found growing on the trunks of trees as well as in the crowns with other epiphytes.

Extremely dainty and delicate plants are the Hymenophyllaceae which, because of the transparency of their leaves (composed of only a single layer of cells), are called filmy ferns. Members of the genera *Trichomanes* and *Hymenophyllum* often grow on treetrunks in tropical mist forests where they might easily be mistaken for liverworts or mosses. Much more conspicuous on trees are the striking tufts of spleenworts *(Asplenium)* (Fig. 438), staghorn ferns *(Platycerium)* and drynarias *(Drynaria)* (Fig. 440). Members of the last two genera form unusual forked leaves in which humus is accumulated for the spreading roots of the plant. Ferns of the tropical forests also include robust climbing forms *(Lomariopsis).* A typical fern of mangrove woodlands is *Acrostichum aureum,* which tolerates not only regular submersion at high tide, but also the resulting increase in salinity, and oxygen deficiency in the soil.

Flowering plants

The most numerous and successful of presentday forest plants are gymnosperms (mostly coniferous trees mentioned in the preceding chapter) and angiosperms (flowering plants). The latter include both the monocotyledons—plants which owe their name to the presence of only a single seed-leaf but are most readily recognized by their parallel veined leaves—and dicotyledons which have two seed-leaves and net-veined leaves. Of the

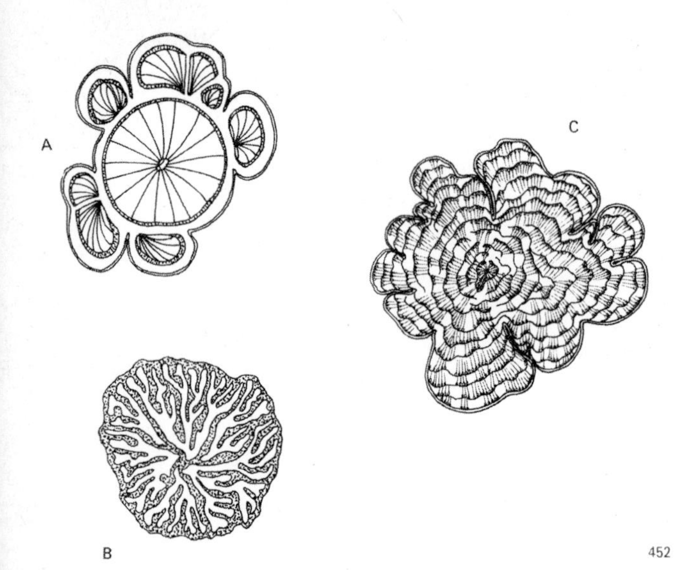

452 Section of the stem of woody lianas with abnormal secondary thickening:
A — *Combretum* sp.
B — *Neuropeltis prevosteoides*
C — *Tiliacora macrophylla.*

453 The strong loops of the liana *Stigmaphyllon sagraeanum* cut into the trunk of a Live Oak *(Quercus virginiana);* Cuba.

454 Clumps of Mistletoe *(Viscum album)* on a White Poplar *(Populus alba)* look like decorative ornaments on a Christmas tree.

total number of 350,000 species of green plants eighty per cent, in other words 280,000, are angiosperms and of these dicotyledons outnumber monocotyledons by three to one. Of course angiosperms include the tree palms (monocotyledons) and broadleaved trees (dicotyledons) described in the preceding chapter but these represent only a small fraction of the angiosperms occurring in various forms in forests from the polar timberline to the equator. Overall forest flora includes more than fifty per cent of all existing angiosperms—that is some 150,000 species. These however are unequally distributed throughout the world. Extreme environments and catastrophic changes in the past are responsible for the fact that the forest flora of some regions numbers only several hundred species (Scandinavia, Canada) and elsewhere tens of thousands (The Amazon Basin and Malaysia).

From this great wealth it is only possible to select a few examples of flowering herbaceous plants. Many have ornamental or variegated foliage and decorative flowers or fruits that brighten the forest (Fig. 441). They may appear to be a supplementary feature but in reality they are far more than mere decoration. In the forest ecosystem they play an important role as producers of green matter, as a factor influencing the composition of humus, as food for forest animals and as competitors of trees—particularly seedlings—for food, water and light. For botanists and foresters, moreover, such plants serve as indicators of environmental factors and forest types (Fig. 443).

Forest species of the buttercup (Ranunculaceae) and primrose (Primulaceae) families all have striking white, yellow, red or blue flowers that are pollinated by forest insects. Figs of the Mulberry family (Moraceae) are important to tropical forests. There are about 800 species including a great many trees as well as numerous shrubs, lianas, epiphytes and stranglers (Fig. 442). Stranglers usually start out in life as epiphytes in the crowns of trees (Fig. 444). Later they send aerial roots down the trunk of the host tree to the ground. These aerial roots branch, thicken and join to form a continuous mantle that strangles the host tree, preventing access of sufficient oxygen as well as restricting growth in girth. The crown of the strangler also branches to form a separate structure in place of the crown of the host tree. The dead trunk of the host gradually de-

455

456

455 Flowering *Dissotis entii.*

456 Detail of flowering *Preussielia chevalieri* which grows as an epiphyte on the branches of tropical trees.

457 Striking is the foliage of *Gunnera chilensis* in moist American forests; mountain mist forest, Ecuador.

composes and in time the strangler fig stands alone like a massive hollow pillar.

The structure of the woody stems of lianas is peculiar. They are frequently square in cross-section, and are flattened, fluted, twisted or seemingly composed of strands resembling those of a cable (Fig. 453). The irregularities of the surface are caused by the unusual way in which the stem thickens (Fig. 452). The stems of woody lianas do not increase in thickness through the action of a single cambium, which in the common woody plants of Europe and North America regularly produces new phloem and new wood. In older liana stems there are several concentric layers of locally circumscribed centres of cambium activity, which is why on a cross-section of a liana stem the new wood and new phloem appear as alternating concentric circles or as surface outgrowths or scattered spots. Woody lianas have large cellular tubes through which water containing dissolved nutrients is carried high up to the canopy of the tropical forests.

The forests of the temperate regions do not have many lianas. They are represented in Europe by the Common Hop *(Humulus lupulus)* (Fig. 445 and 446) and in Japan and Manchuria by the Japanese Hop *(Humulus japonicus)* of the hemp family (Cannabaceae). Hops are lianas that twine clockwise or anticlockwise up the trunks of trees. Other types of lianas include leaners (which have no special devices for holding on to a support and simply lean against the trunk), root lianas (which hold on to the support by means of aerial roots), and tendril lianas (which hold on to the support by means of modified leaves or stems called tendrils). Related to the hops is the Stinging Nettle *(Urtica dioica),* which in European forests is a reliable indicator of fertile, nitrogen-

457

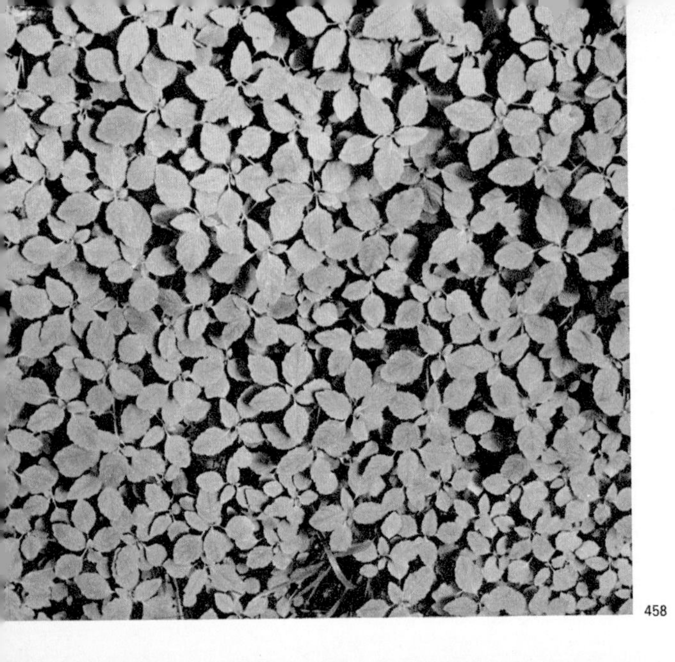

458

458 Growth of young balsam plants Touch-me-not *(Impatiens noli-tangere).*

459 *Impatiens obanensis* is a pink-flowered species growing in the rainforests of Africa.

460 The balsam Touch-me-not *(Impatiens noli-tangere)* makes a thick carpet on the floor of a beechwood.

459

460

461

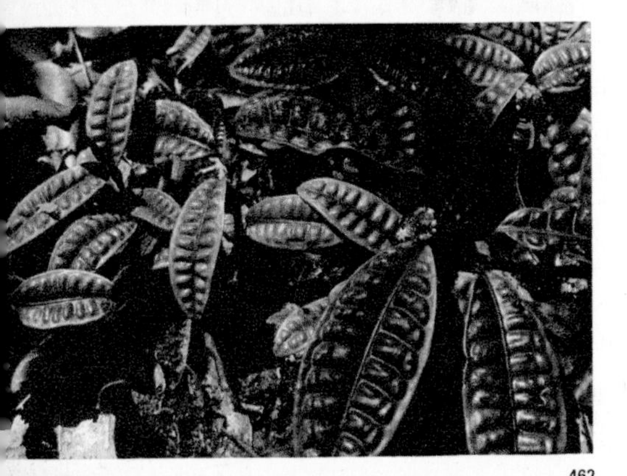

462

463

rich soil (Fig. 445). In all, there are about forty species of nettles, most of which are notorious for their stinging hairs which make life unpleasant for both man and forest wildlife.

Plants of the order Ericales are widespread mainly in the forests of cold and temperate regions. They include, for instance, the Pirolaceae family, green herbaceous perennials that form endomycorrhizae together with forest fungi (Fig. 447), and the shrubs and shrublets of the heath family which are often the dominating element in the forest undergrowth. Distinctive plants of mountain forests and in particular the Himalayas are rhododendrons, whose lovely blooms also ornament suburban parks and gardens where they are grown as cultivated plants (Fig. 448). Related to the rhododendrons are the wild rosemaries *(Ledum)* (Fig. 449) which contain many valuable substances such as volatile oils and tannins and which grow on peat soils in the forests of North America and Eurasia. Associated with these plants is a very distinctive group of animal consumers, chiefly lepidoptera and beetles. Two other groups that must not be overlooked are the heaths *(Erica)*, distributed mainly in South Africa, and the blueberries *(Vaccinium)* (Fig. 451).

The daisy family (Compositae), one of the largest among the angiosperms, is not as abundant in forests as in the open savannas, steppes, tundras and high mountains, but it is usually found at least in forest clearings where it is represented by such plants as groundsel *(Senecio)*, agrimony *(Eupatorium)* and golden rod *(Solidago)*. In Europe hawkweeds *(Hieracium)* are commonly found in forests with acidic humus, whereas the damp soil of submontane forests is best for *Adenostyles alliariae* and *Petasites albus* (Fig. 450). In North American forests composite herbs are even more common.

In forest communities hemiparasitic plants of the Loranthaceae family are frequently encountered, growing on branches in the crown or occasionally on the trunks of a host tree. European forests contain only a few species of mistletoe *(Viscum)* (Fig. 454) and *Loranthus*, but there is great diversity in this group in tropical forests. In fact, the Loranthaceae include 1,300 species, and in the forests of Africa a single branch, less than ten centimetres across, may be covered with as many as fifty shrubby partial parasites belonging to five or more species. Often they also include hyperparasites and autoparasites. The first is a case

461 Sweet Woodruff *(Asperula odorata)* grows in abundance in Europe's forests.

462 The sculptured blade of a forest herb of the genus *Cephaelis* in a tropical forest.

463 The winding trunk of a liana of the Convolvulaceae family in an African forest.

464 *Streptosolen jamesonii* is one of the members of the nightshade (Solanaceae) family growing at lower altitudes in the mountains of Ecuador and Colombia.

465 The yellow-throated white flowers of *Lankesteria brevior* in an African rain forest.

464

465

of a parasite living on a partial parasite of a different species and obtaining its nourishment from the host tree through the partial parasite. Some species of Loranthaceae are clearly hyperparasites that occur on specific species of hemiparasitic relatives. In the case of autoparasitism an individual of one species is attacked by a younger individual of the same species. All parasitic members of Loranthaceae obtain water and nutrients from their hosts, unlike the many epiphytes high up in the treetops of the tropical forest which may attain the size of a large shrub without markedly affecting the metabolism of the tree. One example is *Preussielia chevalieri* of the Melastomataceae family (Fig. 456). Many members

466

467

of this family grow in the herb layer where some species produce ornamental blooms even in deep shade.

When summer is at its height European and North American forest clearings are flooded with the flowers of the Rosebay Willowherb or Fireweed *(Chamaenerion angustifolium)* of the Onagraceae family. The violet-red blossoms are a common sight in the colder regions of the temperate zone. Related willowherbs *(Epilobium)* are more likely to be found in forest swamps. Growing in similar habitats but in the rainforests and mountain mist forests of the tropics are the huge herbs of the Gunneraceae family (Fig. 457), limited in occurrence to the southern hemisphere.

Fresh soil rich in humus is required by the balsams *(Impatiens).* There is an amazing similarity between the Asian and African species (Fig. 459). All balsams produce pods which burst at the slightest touch scattering seeds in all directions. The seeds have excellent powers of germination and produce in damp forests, broad carpets of seedlings (Fig. 458) and thick masses of adult plants (Fig. 460).

Most members of the madder family (Rubiaceae) are tropical plants. They include both woody and herbaceous plants which are also in central and north European as well as Canadian forests. *Mussaenda erythrophylla* of Africa's forests is an example of a leaning liana

466 The flowers of the Yellow Star of Bethlehem *(Gagea lutea)* rise from the dry forest litter in spring.

467 The Toothwort *(Lathraea squammaria)* is a parasite on the roots of shrubs and trees.

468 Herb Paris *(Paris quadrifolia)* and Sweet Woodruff *(Asperula odorata).*

469 The May Lily *(Majanthemum bifolium)* is a common sight in European forests.

468

469

or shrub. It has enlarged, strikingly red sepals that glow brightly in the dimness of the forest. In Ghana it is known as Ashanti Blood. In some madders of the shady tropical forests the leaf blades are of an unusual shape (Fig. 462) which allows them to dry quickly during long rainy spells. The leaves and leaf-like appendages at the base of the leafstalks (stipules) of European herbs of the same family are far less striking (bedstraw—*Galium* and woodruff—*Asperula),* but even so they are very typical forest plants (Fig. 461).

Many of the 1,000 species of the morning-glory or bindweed family (Convolvulaceae),

470

470 Dracaenas of all sizes thrive in tropical forests; the picture shows *Dracaena adamii.*

471 The widely distributed *Gloriosa superba* is a native of the forests of Africa and Asia.

471

particularly the numerous ipomoeas, are also found in the forest undergrowth, though the most extraordinary are the long thick lianas of this family. *Neuropeltis prevosteoides* of the African rain forests (Fig. 463) probably has the strongest stems and most complex structure of all.

Larger yet is the nightshade family (Solanaceae), found in vast numbers in all tropical forests (Fig. 464). Deadly Nightshade *(Atropa belladonna)* is also well known in damp European forests, as is Woody Nightshade or Bittersweet *(Solanum dulcamara).* They all contain alkaloids used by man both for his benefit and to his detriment.

The herbs and shrubs of the forests of warm

The Slipper Orchid *(Cypripedium calceolus)* has a wide area of distribution.

The ornamental foliage of *Caladium* stands out in the undergrowth of the tropical forest; Brazil.

Flowering *Heliconia* on the edge of the rainforest; Ecuador.

472 The decorative *Haemanthus cinnabarinus* grows in the shade of Africa's rain forests.

472

473 The white flowers of *Crinum jagus* of the Amarillidacea family glow brightly in the dimness of the tropical forest.

473

474 The broadleaved *Mapania coriandrum* of the sedge family grows in the shade of the tropical forest.

475 The only member of the genus *Eurychone rothschildiana* growing on the trunk of a tree in virgin forest in Africa.

474

and tropical regions also include many members of the Acanthaceae family (Fig. 465). In African forests alone there are some fifty genera of this family and many are the dominant plants beneath the canopy.

Angiosperms also include parasitic plants (described in previous chapters). Such plants are unable to produce their own food and are thus dependent on their host, for all their mineral and organic nutrients. These totally parasitic plants are to be found amongst the Rafflesiaceae, Hydnoraceae, Balanophoraceae and Orobanchaceae families. Whereas *Rafflesia* is the primadonna of the entire plant realm, the modest toothworts, (Fig. 467), or various members of the genus *Orobanche* are barely visible parasites among the forest undergrowth.

Monocotyledonous angiosperms (Monocotyledonae) are often the dominant plants of the forest, giving the whole its specific character—as, for example. grasses (Gramineae), sedges (Cyperaceae), aroids (Araceae) and palms (Arecaceae). Man, however, delights most in the presence of numerous members of the lily (Liliaceae) and orchid (Orchidaceae) families, which have unusual and attractive flowers as well as leafy stems.

475

476

It is hard to imagine spring in the forest of Canada and the northern United States without trilliums and erythroniums and the central European woods without the Yellow Star-of-Bethlehem *(Gagea lutea)* (Fig. 466), Solomon's seals *(Polygonatum)*, lilies *(Lilium)*, *Allium* species, Herb Paris *(Paris quadrifolia)* (Fig. 468), May Lilies *(Maianthemum bifolium)* (Fig. 469) and the aristocratic Common White-hellebore *(Veratrum album)* (Fig. 418).

Lilies are perennials which survive the vagaries of winter in temperate regions as bulbs, tubers or rhizomes. These organs have proved to be very efficient forms, but currently they are resulting in the annihilation of many lily populations, for bulbs are being dug up and transferred to suburban gardens (with varying success). Many lilies grown as house plants are of forest origin. Dracaenas, for instance, are striking herbs, shrubs or small trees in the forests of Asia, Africa and America (Fig. 470). Some, particularly *Dracaena fragrans,* are a decorative feature in many European homes as are the five species of *Gloriosa,* herbaceous lianas which climb with the aid of tendrils at the tips of the leaves, and which bear lily-like flowers of inimitable beauty (Fig. 471).

477

476 *Microcoelia caespitosa* in full bloom; the vegetative organs are mainly the green roots where photosynthesis takes place.

477 *Cyrtorchis arcuata* is a fragrant species growing throughout equatorial Africa.

480 *Cymbidium lowianum* in the mountain forests of the Himalayas.

478 The orchid *Epidendron vespa* growing in a gap in a mountain mist forest in Ecuador.

479 *Cephalantera alba* is a ground orchid of Europe's broadleaved forests.

Another family of monocotyledons not too far removed from the lilies is the Amaryllidaceae which also includes very pretty flowers that grow in temperate and tropical forests. In European forests that have humus-rich soil and are rather damp, spring is heralded by the Snowdrop *(Galanthus nivalis)* and Spring Snowflake *(Leucojum vernum)* while the trees are still bare. To the city dweller who has lost contact with nature's calendar they are a welcome reminder that spring is just around the corner.

Tropical amaryllids are far more impressive, of course, particularly the various species of *Haemanthus* (Fig. 472) and *Crinum* (Fig. 473).

Members of the Dioscoreaceae family, which include twining lianas and small herbaceous plants of both the wet and dry tropics, have a somewhat unusual position among the monocotyledons. Many species of the genus *Dioscorea* have starchy tubers (yams) that are a staple food locally; elsewhere they are grown as ornamental house plants.

The deep shade of a closed forest is generally not ideal for the growth of sedges, rushes (Juncaceae) and grasses. These families do best in the full sun of tundras, steppes, open

480

481 Plants growing at the edge of a tropical rainforest in Ecuador include *Philodendron verrucosum, Anthurium hygrophilum* and members of the Melastomataceae and Myristicaceae families.

482 The screw-pine *Pandanus candelabrum* in the swamps of the Atewa Range, Ghana.

savanna woodlands and swamps. In the cooler forests of Eurasia and North America there are usually only scattered tufts of sedge *(Carex)*. Occasionally, as a result of man's activities, there may be large spreading masses of these plants. *Carex brizoides* often forms dense masses in forests subjected to heavy grazing by cattle. The narrow-leaved grass-like life forms are practically never found in the shade of the rainforest though some broadleaved members of the sedge family may grow there (Fig. 474).

Noted for the number of species, and above all for its remarkable blossoms, is the family of orchids. Many people travel to the tropical forests only to see or collect the beautiful orchids growing there. They are true orchid hunters and their lot is no easier than that of those who hunt exotic game. Tropical orchids grow most luxuriantly in the treetops (Fig. 475, 476 and 477) where they form whole gardens

483 The rotang palm *Ancistrophyllum secundiflorum* climbs up into the treetops in a rainforest in Ghana.

484 New growth of seedlings and young solitary palms in a swamp forest in Ghana.

483

484

485

beyond the reach of human hands and often also out of sight because the dense canopy makes it impossible to see the uppermost layer from below. Furthermore, individual orchids flower intermittently and all species do not flower at the same time. Flowering specimens occur only here and there and tufts of leaves are hard to identify by themselves without the flowers. It is difficult to distinguish between the 1,000 species of *Bulbophyllum* when they are in bloom, let alone without the flowers.

Besides epiphytic species, orchids include small lianas, and many types rooted in soil (with an endomycorrhiza). The Slipper Orchid *(Cypripedium calceolus)* is known to every nature lover in the United States, Canada and Europe. Also well known are many forest orchids of the genera *Orchis, Platanthera, Listera, Epipactis* and *Cephalanthera* (Fig. 479). Much more impressive in size and shape of flowers, are the ground orchids of the tropics (Fig. 478). Even in the mountain forests of the

486

485 The epiphytic *Tillandsia fasciculata* on the trunk of *Quercus sagracanum* is partly covered by a newly built termite nest; Cuba.

486 The short *Sclerosperma mannii* palm in a swamp in the tropical forest of Africa.

487

Himalayas there grows an orchid which is an additional reward to the mountain climber who scales the heights (Fig. 480).

Moist coastal forests as well as inland forests are inhabited by shrubs and small trees of the screw-pine family (Pandanaceae). Striking characteristics are the aerial roots and the crowns of narrow leaves with serrated edges (Fig. 482). Only a general change in the environment will expel screw-pine from its established site. Other very distinctive plants of tropical forests are the aroids (Araceae). They are mostly herbaceous, rarely woody, plants, growing on the ground or as epiphytes in soil pockets in the treetops. Many species climb up treetrunks like small root lianas. The aroids are noted for the diversity of their leaves, for which reason they have become perhaps the most popular house plants in the world. No modern home is without its philodendron, its monstera or its caladium, though no such arrangement can rival the magnificent natural tropical display.

Palms of the Arecaceae family have been mentioned several times in the preceding chapters. They may be trees with a bole up to fifty metres high but there are also many smaller, branching types that could easily be

488

487 *Marantochloa purpurea* grows to a height of three meters in waterlogged soil in the heart of an African rain forest.

488 The ground in the tropical forest is covered by *Costus englerianus.*

489 Bamboo (*Bambusa* sp.) in Japan.

490 The open tropical forests of Africa are invaded by the grass *Setaria chevalieri.*

491 *Guaduella oblonga* is a primitive broadleaved grass found in the undergrowth of African forests.

490

491

called shrubs, as well as many leaning lianas and types that are best classed as perennial herbs. On the floor of the virgin rain forest are found solitary leaves with slender stalks and only slightly divided blades. These are usually the small seedlings of future lianas or trees (Fig. 484). Passage through the forest is hampered by these lianas, collectively called rattan palms of the genus *Calamus* and allied genera. A rattan palm climbs by means of a stem that adheres to its support by means of sharp spines on the underside of the leaf rachis or by means of spiny tendrils at the tip of the rachis (Fig. 483).

Also remarkable are palms with undivided leaf blades (Fig. 486) which grow like huge fans to a height of several metres. As has already been pointed out, palm trees possess the largest leaves of all terrestrial plants.

In the forests of tropical America the com-

492 A mountain spruce forest with a continuous carpet of *Calamagrostis villosa.*

monest epiphytes are bromeliads. No one knows why this vigorous group has not spread to other continents. Only one other species occurs elsewhere—on the coastal cliffs of Senegal. The flowers and inflorescences of bromeliads exhibit less diversity in shape and colouring than those of orchids, but a decorative feature of these plants is the leaves, which are various shades of green, red and blue with interesting markings. Bromeliad seedlings root readily in cracks in the bark and then it is only a matter of time before they grow into a sizeable clump (Fig. 485).

Robust herbs growing below the canopy of tropical forests are also to be found among the members of the Strelitziaceae, Zingiberaceae

492

(Fig. 488) and Marantaceae (Fig. 487) families. In low-lying spots with waterlogged soil the dense canopy is often broken because such spots are covered with masses of Marantaceae that, like reeds, make it impossible for trees to gain a foothold.

Narrow-leaved grasses are not to be found in shady tropical forests, but growing there to this day (and mainly in south-east Asia) are their predecessors, giant bamboos. In many instances they are not mere undergrowth but form an independent biome in which they are the dominating plants (Fig. 489). Some, such as species of the genera *Dendrocalamus* or *Arundinaria,* grow amazingly quickly to heights of up to thirty-five metres. Also found in the shade of tropical forests are early forms of grasses with broad leaves (Fig. 491).

In forest clearings throughout the world and wherever conditions do not favour forest development, or where forest passes into savanna, steppe or tundra, narrow-leaved grasses (Fig. 490 and 492) will be found. Here, growing in juxtaposition, are the two most vigorous life forms of the plant kingdom. On the one hand are the more productive and efficient woody plants and on the other the definitely hardier and less demanding grasses. Advances and retreats of the line of demarcation between the two occur in response to natural conditions over a period of many years, and as a result of the influence of man.

493

THE ANIMALS OF THE FOREST

The forest is full of animals which no zoo in the world can rival in number and variety. Even the poorest forest is the home of many thousands of species belonging to many different animal groups. Universally, animals far outnumber plants, even though the latter include microscopic bacteria and fungi which some biologists class in separate kingdoms. To date the number of animals described by zoologists exceeds the number of described plants by three to one and it is estimated that when most species of plant and animal life have been described, animals will number five to ten million species, which will be five to ten times that of all known species of plants. More than half are denizens of the forest, for this is the chief habitat of the largest group of animals, the insects. In other words, several million species of animals are adapted to life in the forest—in the treetops, on and inside treetrunks, on flowers and fruits, in carpets of moss, in humus, around roots in the soil, or living as parasites on other forest animals.

Compared with other terrestrial biomes the forest provides animals with far more living space both above and below the ground. Between the upper treetop level and the bottom layer of pioneer roots there is plenty of room for the large populations of both invertebrate and vertebrate species. The forest environment is characterized by great diversity with varying availability of light, heat, water and food. Chiefly responsible for this diversity are the forest plants and animals themselves. The leaves, flowers, fruits, seeds, buds, wood, phloem, bark and roots of the different species vary in compositon and structure and provide an immense variety of food for animals as well as places for them to hide. During the course of evolution the forms of forest animals and plants became mutually adapted. Important to the existence of a great number and variety of forest animals are animals themselves, for they are the food of predators, parasites and saprophytes.

493 A Yellow-shafted Flicker *(Colaptes auratus)* of North America.

Protozoans

Moss, humus, soil, forest puddles and also the skin and mucous membranes of forest animals are inhabited by protozoans—mostly microscopic animals each consisting of a single cell with a variously formed enclosing membrane or case. Common in forest soil are relatives of the well-known *Amoeba* as well as forms which are encased in a 'shell', such as the saprophytes *Arcella* and *Euglypha* which live on organic material from dead plants in humus. Large numbers of protozoans live in the digestive tract, muscles and blood of insects, birds and mammals. Some live together with their host in a relationship that is mutually beneficial or at least unharmful (symbiosis), while others are dangerous parasites and cause serious diseases. Parasitic protozoans find optimum conditions for growth chiefly in tropical forests where they attack practically all species of vertebrates, including man, unless he protects himself by using modern preventive measures. They include dangerous flagellates of the genus *Trypanosoma* (the cause of sleeping sickness) and *Leishmania,* relatives of amoeba including the Dysentery Amoeba *(Entamoeba histolytica),* and species of *Plasmodium,* the cause of malaria. Many species of protozoans require two different hosts for their development and two different environments. For example, plasmodia live in the red blood corpuscles of vertebrates and in the alimentary canal and salivary glands of mosquitoes which suck the blood of vertebrates. The place and time of food and energy transfer from one member of the forest ecosystem to another are often the place and time of infection.

Worms

Living freely in the soil and as parasites in plants and most animals are various worms: flatworms (Plathelminthes), aschelminths (Aschelminthes) and annelids (Annelida) (Fig. 494). Forest game is often infested by flukes. The liver fluke *(Fasciola hepatica)* requires an intermediate host, the water snail *(Limnaea truncatula),* to complete its complex life cycle. Aschelminths are represented in the wet for-

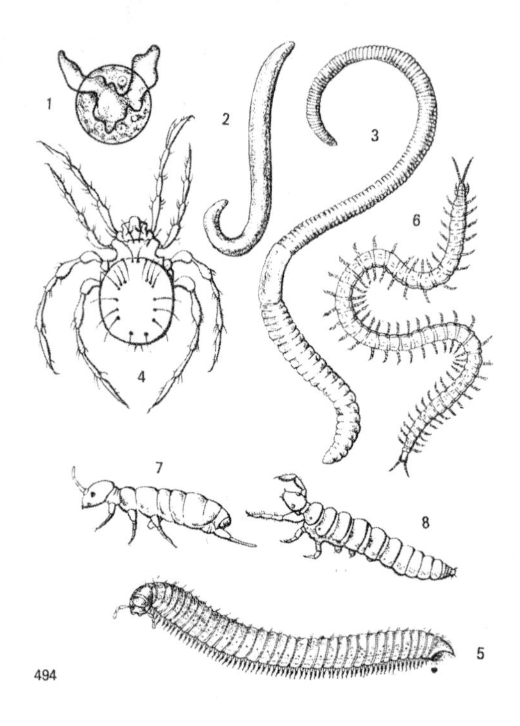

est soil by rotifers (Rotatoria), and numerous roundworms (Nematoda) live in humus, on roots, and in the bodies of forest animals. Most important of the annelids are the earthworms (Oligochaeta) and the leeches (Hirudinea). Where conditions are favourable earthworms are so numerous that they represent the largest part of the animal biomass. They draw large amounts of humus-rich soil through their alimentary canals, eliminating the undigested remains in the form of tiny round crumbs or larger mounds and thus aerating and fertilizing the soil. It is impossible to imagine tropical forests without leeches, which hang on herbs and trees and attach themselves to any animal, including man, which brushes past them. It is surprising that the African forests lack these worms.

Molluscs

The forest environment is apparently not very congenial to molluscs, which is the second largest phylum in the animal kingdom after the arthropods. Nevertheless, the forest harbours at least one group—namely the gastropods (snails and slugs). Snails of the genus *Achatina* have a shell that is fifteen centimetres long and it is a delight to watch them climb slowly over fallen trees in the tropical rainforest. *Arianta arbustorum* and the Roman or Edible Snail *(Helix pomatia)* of European forests have far smaller shells and slugs of the Limacidae and Arionidae families, commonly found in damp forests in temperate regions (Fig. 495), have no shells whatsoever.

Arthropods

Forests are the home of numerous arthropods—spiders (Arachnida), crustaceans (Crustacea), millipedes (Diplopoda), centipedes (Chilopoda) and insects (Insecta). Widely distributed in the forests of warm regions are the scorpions, which hunt smaller animal life. Many a man has been the victim of their poisonous sting and some physicians working in the tropics assert that the number of deaths caused by snakebite is far smaller than the number caused by the sting of poisonous scorpions. Spiders are found in forest areas; some are very large and even hunt small vertebrates. Other species are web spinners and their webs may be found near the ground and at higher levels in the forest (Fig. 496 and 497).

A great many species of ticks and mites

494 Microscopic and small animals living in forest soil:
1—protozoan, 2—aschelminth, 3—annelid, 4—mite, 5—millipede, 6—centipede, 7—springtail, 8—telsontail of the genus *Eosentomon*.

495 The Large Red Slug *(Arion rufus)*.

A snail of the genus *Achatina* of the tropical rainforest.

The Pine Hawkmoth *(Sphinx pinastri)* on the bark of a Scots Pine.

The larva of the Pine Hawkmoth *(Sphinx pinastri)* feeds on pine needles.

The butterfly *Agraulis vanillae* in the undergrowth of an Amazonian forest.

496

497

496 A garden spider of the genus *Argyope* in the centre of a web in a virgin forest in Africa.

497 The morning dew and slanting rays of the sun make the web of a forest spider a thing of beauty.

498 An isopod of the family Oniscidae.

499 Female Sheep Tick *(Ixodes ricinus)* gorged to the full.

500 The millipede *Polydesmus complanatus* under the bark of an old stump.

(Acarina) live in the ground, where they play an important role in the decomposition of forest litter and the formation of humus. Other species are parasitic on plants and animals. The mixed and broadleaved forests of the warmer parts of Europe are the home of the Sheep or Castor Bean Tick *(Ixodes ricinus).* Its early larvae are parasitic on small forest mammals and birds, the later stages on medium-sized mammals and birds, and the adults on large mammals, including man. A gorged female sheep tick (Fig. 499) in the hair of a forest ungulate is not at all uncommon. Besides sucking the blood of its victim it may also transmit a variety of diseases.

Crustaceans of all sizes live in the soil in forests. The mangrove woodlands and freshwater swamp forests of the tropics are commonly inhabited by crabs (Brachyura) which excavate large corridors in the soil thus improving drainage and the access of air to the plant roots. In the soil of drier forests much smaller crustaceans, such as woodlice (Oniscidae) (Fig. 498), consume plant remnants.

Millipedes are also primarily herbivorous or feed on decaying plant remains in the soil. Many centipedes, on the other hand, are predaceous and feed on small animal life within the soil.

As regards multiplicity and diversity of species and adaptability to widely different environments, none of the aforementioned groups of arthropods can rival the insects. They are

501

501 The Praying Mantis *(Mantis religiosa)* has the front pair of legs specially adapted for catching prey.

502

502 A locust (Caelifera) of the dry forests of the Mediterranean region.

503 A cicada (Cicadidae) of the virgin forests of New Guinea.

503

found in the dark depths of the soil as well as high up in the treetops and include species limited to specific foods and species which are more catholic in their feeding habits.

Many species are closely associated with a specific species of forest plant or animal as regards their food, habitat or territory.

Found in forest litter and in the ground is a group of wingless insects (Apterygota), which includes various springtails (Collembola), telsontails (Protura) and two-pronged bristletails (Diplura), that crush and decompose plant remnants and without which many forest ecosystems would be unable to continue the smooth cycling of nutrients. Far more active and versatile are the winged insects (Pterygota), most of which possess two pairs of wings, whose performance not even modern technology can equal. The method of locomotion, the sense organs, mouth organs and the method of reproduction of these insects are adapted to the immense depth and division of space in the forest ecosystem. Some species have even become incapable of flight, or lack wings altogether.

Of great importance in the life of the forest are the termites (Isoptera). Many tropical species consume dead wood, others feed on dead plant stems and leaves in the forest litter, while others obtain nourishment from humus-rich soil, depleting it of nutrients. They build impressive nests in the forests (Fig. 501). These nests are also inhabited by great numbers of organisms, such as fungi of the genus *Termitomyces* and other fungi, on which the termites feed, and by numerous insects, some of which live peacefully with the termites while others are predaceous and attack their hosts. Every termite nest is thus an independent community with very complex interrelationships.

Termites that construct their nests in the canopy of the tropical forest build and maintain a system of corridors on the surface of the treetrunk leading from the ground to the top of the tree. These connecting corridors are made of clay and are located on the surface of trunks and branches and sometimes also on the stalks of large leaves. By this means termites have constant access to their source of food, water and building material, and by this means large quantities of soil, which, together with associated humus, may become a fertile substrate promoting the growth of numerous epiphytic plants on treetrunks and in the treetops. Large amounts of soil are also transported to the crowns by termites that live in underground termite nests or in nests on the ground. Many such species gather food on herbs and woody plants and because they cannot tolerate light, they build connecting corridors covered with cemented clay which is washed away by the first heavy rainfall thus forcing the termites to begin building all over again. In this way soil from the deeper layers is brought to the surface, where it mixes with the layer of humus.

Many termite nests are small structures. For example, one East African forest species builds a nest that looks very much like the stinkhorn fungus in shape and size.

The digestive systems of termites which feed exclusively on wood have become adapted so that they can process food that contains little readily available nourishment. Their gut

504 The phasmid (Phasmida) blends perfectly with the forest twigs.

504

505

506

harbours large numbers of flagellate protozoans which partially digest the cellulose, and the termites then use the sugar thus produced. Young termites feed initially on the salivary secretions of the worker termites and only after the symbiotic flagellates have invaded their digestive tracts do they begin feeding on wood.

Long-horned grasshoppers (Ensifera) and locusts or short-horned grasshoppers (Caelifera) (Fig. 502) are very numerous in the forests of warmer regions. The former are mostly carnivores, feeding primarily on insects, and are therefore useful, whereas locusts, which often migrate in swarms in the tropics and subtropics, are herbivores and frequently cause grave damage resulting in marked changes in the vegetation. The well-known Migratory Locust *(Locusta migratoria)* which devastates dry savanna woodlands, also occurs in rain forests where small swarms cause feeding damage in the treetops throughout the year. Some specialized species may be heard at dusk, particularly round flowering trees or trees with newly emerging leaves. Further typical inhabitants of tropical forests are stick insects and leaf insects, or phasmids (Phasmida), which resemble dry sticks (Fig. 504) or green leaves. Many true bugs (Heteroptera) merge into their surroundings in a similar way. The Green Shield Bug *(Palomena prasina),* which feeds on the sweet juices of forest fruits, leaves in its wake an unpleasant smell which has destroyed the appetite of many a blueberry or raspberry picker.

Forests are also full of homopterous insects (Homoptera). Cicadas (Cicadinea) (Fig. 503), aphids (Aphidinea) and mealy bugs (Coccinea) are widely distributed in the temperate and tropical regions. Aphids, in particular, multiply in large numbers under favourable conditions, and exert a considerable influence upon the development of forest plants, by sucking their sap, causing their leaves to deform, dry up or become covered with galls, and by transmitting virus diseases. Aphids have many enemies amongst insects and birds and thus population explosions of aphids are followed by population increases in other forest organisms, such as the Seven-spot Ladybird *(Coccinella septempunctata).*

The complexity of life in the forest ecosystem is greatly increased by the presence of the hymenopterous insects (Hymenoptera) which

505 The ichneumon fly *Rhyssa persuasoria* laying eggs.

506 *Tachyptera atrata* laying eggs in a trunk.

507 The pagoda-shaped nest of termites of the *Cubitermes* group in the undergrowth of the tropical forest.

508 The Wood Ant *(Formica rufa)*.

507

508

509

510

511

include a vast number of species—several thousand in central European forests alone. In the various stages of their development they feed on the leaves of trees and destroy the wood. Others are content with the sweet nectar and pollen of forest blossoms, at the same time playing an important role as pollinators. Still others are parasitic on other insects and help maintain the stability of the forest ecosystem. Pine needles, for example, are consumed in large quantities by the larvae of web spinning and leaf rolling sawflies (Pamphilidae and Diprionidae). Characteristic galls are made on plants by the gall wasps (Cynipidae). The various groups of wasps—Vespidae, Masaridae, Pompilidae, Sphecidae—are herbivorous as well as carnivorous. Ichneumon flies (Ichneumonidae) include species that are parasitic on various insect pests (Fig. 505). Hymenopterous insects also include many different species of bees (Apidae) which suck the

509 The ants with a caterpillar.

510 Swarming ants of the species *Lasius fuliginosus.*

511 Swarming Honey Bees *(Apis melifera).*

nectar of flowers and collect plant pollen (Fig. 511). They are important pollinators of forest plants in the temperate and tropical regions.

The life of a forest is greatly influenced by the activity of ants (Formicoidea) (Fig. 508, 509, 510). Not only are they important because of the wide range of food they eat—their numbers include herbivores, carnivores and omnivores—but also because of their large populations and their extensive building activity. Between ants, plants and some animals there exists a more or less permanent relationship referred to as myrmecophily. The precise nature of these relationships has for years been the subject of research by zoologists and botanists who have obtained much interesting material.

The most clear-cut example of such an association between ants and trees is that of ants of the genus *Azteca* and the cecropia trees of South America. The ants almost always live in the hollow parts of the trunk and protect the cecropias against attack by the leaf-eating ants of the genus *Atta* (Fig. 513). The tree provides the ants with shelter and also with a supply of sweet liquids. Similar close ties exist between American and East African acacias and ants inhabiting gall-like growths at the base of the leaves. In this instance scientists were interested in determin-

512

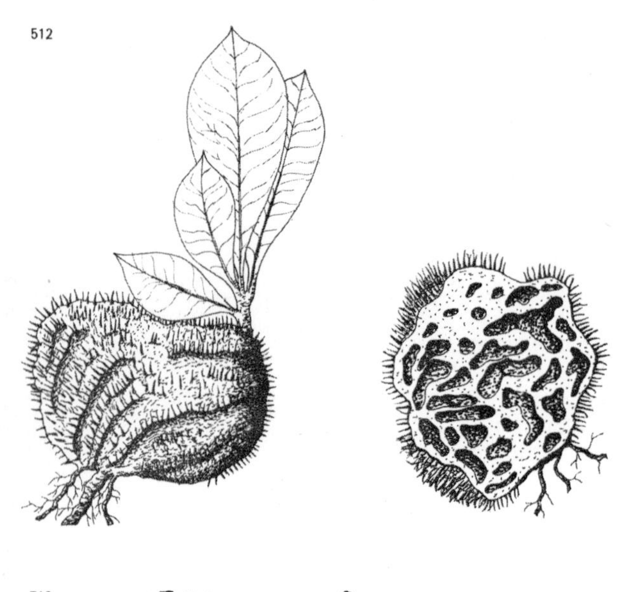

512 The epiphytic plant *Myrmecodia echinata* which is usually inhabited by ants (example of myrmecophily).

513 Activity of the leaf-eating ants *Atta colombica tonsipes* in a tropical forest in Central America; the black triangles mark the path from (A) along which the ants carry pieces of leaves, flowers and fruits to the nest (B), which contains fungus gardens that provide the ants with further food; (C) the black circles mark the path along which the ants carry undigestible remnants and waste matter to a refuse heap (D) or to a stream (E).

513

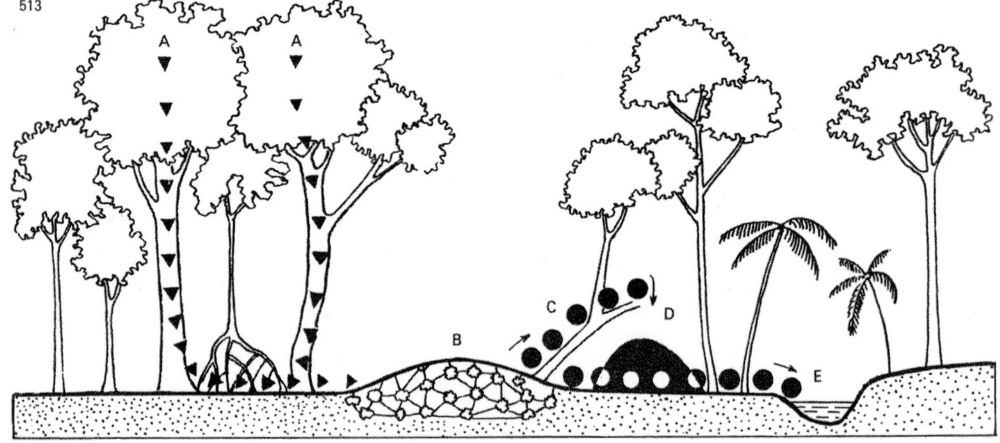

514 The Common Ground Beetle *(Calosoma sycophanta).*

515 *Clytus arietis* of the family of checkered beetles (Cleridae).

514

515

ing whether the gall-like growth was made by the ants, was an adaptation by the tree to promote symbiosis or was an organ in its own right, which was formed and continued to exist without the direct influence of the ants. The last proved to be correct.

Even more unequivocal is the connection between woody plants of the genus *Barteria* and ants of the genus *Pachysima* in Africa. The area of distribution of the two is identical and the ants were always observed on the host tree and never elsewhere. These ants are very aggressive and protect the host tree from attack by other insects.

It was found that many plants in tropical forests have various growths inhabited by ant colonies on their leaves, stems or roots. The palm *Korthalsia echinometra* has a swollen stalk with long spines, in *Dischidia rafflesiana* one of the leaves furls to form a case, and in the well-known *Myrmecodia echinata* the lower part of the stem is swollen like a tuber and riddled with cavities and corridors that are often inhabited by ants (Fig. 512).

516

517

516 *Chrysochroa buqueti* of the family of metallic wood-borers on the islands of south-east Asia.

517 The buprestid *Actenodes costipennis*.

Great diversity in shape is exhibited by the gall-like growths on the underside of the leaves and on the leaf stalks of numerous members of the families Rubiaceae, Sterculiaceae and Melastomataceae. In cultivation, however, it was found that in most instances these growths appeared even on plants that were not inhabited by ants. In all probability, therefore, this is a hereditary adaptation.

Ants are important not only in transporting soil from the substrate high up into the treetops but even more in transporting seeds. In European forests the Wood Ant *(Formica rufa)* is well known as a transporter of oily seeds. One nest was found to contain a total of 36,000 seeds, some of which must have been brought from a distance of seventy metres.

There are about 400,000 species of beetles (Coleoptera), with a great diversity of shape and colour. They are to be found in all parts of the forest, both above and below ground, at all times of the year, and their activities influence many aspects of the forest ecosystem. They are an important link in the flow of energy and the cycling of nutrients. Some species are carnivores, others herbivores or omnivores and often the diet of the larvae differs from that of the adult beetles.

Noted for their striking appearance are the ground beetles, represented in central Europe by the Common Ground Beetle *(Calosoma sy-*

518 The male and female Stag-beetle *(Lucanus cervus)* exhibit marked sexual dimorphism.

519 Male and female longhorn beetles of the species *Dorcus parallelopipedus.*

520 The Blister Beetle *(Meloë proscarabaeus).*

518

519

520

521

521 The longhorn beetle *Steirastoma marmoratum* of the Cerambycidae family lives in the forests of Paraguay.

522 Chafers in European pine forests are represented by *Polyphylla fullo* of the Scarabaeidae family.

523 *Dicranorhina derbyana* of the Scarabaeidae family is native to the virgin forests of West Africa.

cophanta) which consumes large numbers of larvae of the most injurious moths (Fig. 514). The carrion beetles (Silphidae) include some species that feed on carrion and some that even bury small carcasses in the ground. The checkered beetles (Cleridae) include many useful species (Fig. 515). The loveliest beetles of all are the metallic wood-borers or buprestids (Buprestidae). Though there are only several hundred species in the temperate regions, many thousands exist in tropical forests (Fig. 516 and 517). Ladybirds (Coccinellidae) eagerly hunt mealy bugs, aphids and their larvae. Blister beetles (Meloidae), found in forest

522

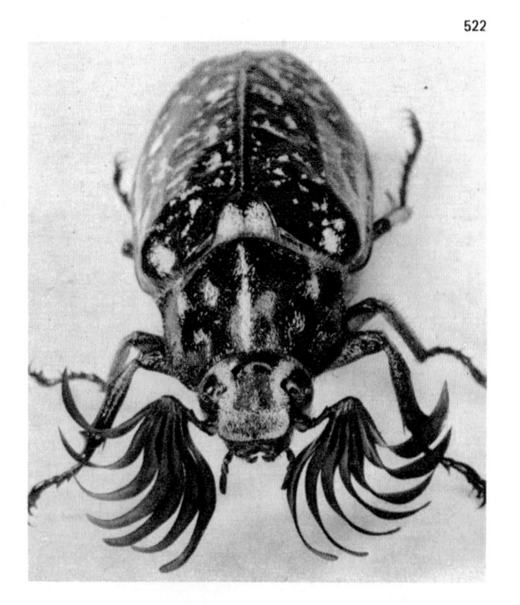

523

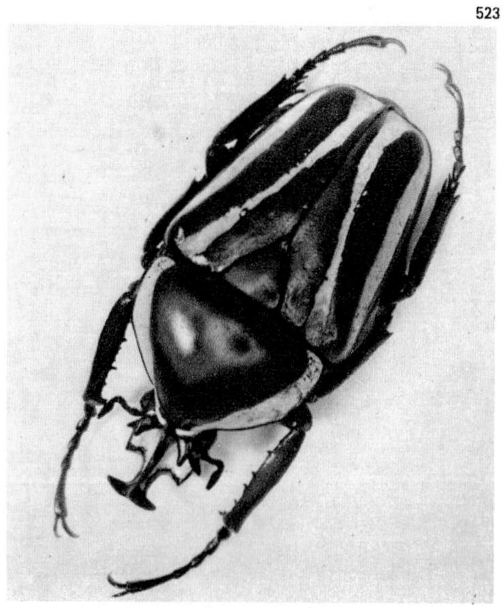

524 The longhorn beetle *Rhagium mordax* of the Cerambycidae family.

525 The Pine Weevil *(Hylobius abietis)* of the Curculionidae family.

524

525

clearings, undergo a complex development, during which they pass through many different stages, one of which takes place in the eggs of solitary bees (Fig. 520). A common inhabitant of Europe's oak and beech woods is the Stag-beetle *(Lucanus cervus)* (Fig. 518), which passes the larval stage in the decaying wood of old trees or stumps. The Scarabaeidae is an outstanding family (Fig. 522 and 523) which includes the well-known cockchafers and chafers, such as the Cockchafer or Maybug *(Melolontha melolontha)*, the Rose Chafer *(Cetonia aurata)*, and the tropical Goliath Beetle *(Goliathus goliathus)*.

Wood makes up the greatest part of the forest biomass and the larvae of longhorned beetles (Cerambycidae) (Fig. 521 and 524) are equipped by nature for its consumption. Signs of their activity may be found in old stumps as well as upright, dead trees. Occasionally they also burrow into living trees and they often devour adult leaf beetles (Chry-

526

527

528

526 A weevil of the genus *Cyrtotrachelus* which lives in the forests of Java.

527 A beetle of the Brentidae family which lives gregariously under the bark of trees in Paraguay.

528 The weevil *Eupholus weiskei* is widespread in New Guinea.

The West Indian Woodpecker *(Centurus superciliaris);* Everglades National Park, U.S.A.

The Great Tit *(Parus major)*.

The young elk *(Alces alces)*.

The European Lynx *(Lynx lynx)* inhabits dense forests as well as mountain and rocky valleys.

somelidae) and their larvae. Beetles of the Brenthidae family (Fig. 527), which are extraordinarily narrow and long, live under the bark of woody plants. The snout-beetles or weevils (Curculionidae) are narrowly specialized and often there may be fifty or more species on a single tree, each adapted for feeding on a different plant organ (Fig. 525, 526 and 528). The larvae of different species of bark beatles (Scolytidae) bore characteristic tunnels on the underside of bark and in phloem, or directly into wood. The adult beetle has a streamlined body that is superbly fitted for passage through the tunnels (Fig. 529, 530 and 531).

In a contest for the most beautiful of all animals butterflies and moths (Lepidoptera) would doubtless be among the winners. They exhibit great diversity in shape but are noted primarily for the fantastic patterning and coloration of their wings. It could be said that nature has achieved the ultimate in artistic design were it not for the fact that the reason behind all this beauty is in many cases quite prosaic—often the insect is camouflaged to match the surroundings and sometimes it is made conspicuous with a variety of shapes and patterns including some which look like eyes. Adult moths and butterflies cause little harm to forest plants but the larvae (caterpillars) of many are avid consumers of plant matter.

The caterpillars of carpenter moths (Cossidae) (Fig. 532) bore tunnels in the wood of forest trees. The forests of North America and northern Eurasia are inhabited by the tortrix moths (Tortricidae), some of which may cause serious damage when they overmultiply, by entirely stripping trees of their leaves. Similar damage is caused by some geometrid moths

529

530

531

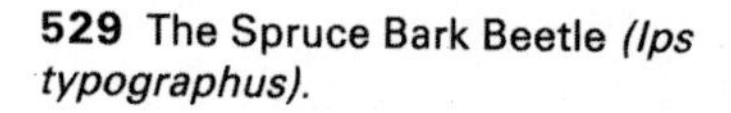

529 The Spruce Bark Beetle *(Ips typographus)*.

530 Larva of the Spruce Bark Beetle.

531 The characteristic feeding pattern made by the Spruce Bark Beetle under the bark of trees is revealed when a piece is peeled off.

532

533

532 The Goat Moth *(Cossus cossus).*

533 A predaceous snake fly (*Raphidia* sp.)

534 The Garden Tiger *(Arctia caja).*

534

535

536

537

538

539

535 Clifden Nonpareil *(Catocala fraxini).*

536 The Emperor Moth *(Saturnia pavonia);* female.

537 A swallowtail *Papilio payeni* (Papilionidae) found in the virgin forests of Borneo.

538 The swallowtail *Papilio euchenor* of New Guinea and the Solomon Islands.

539 Poplar White Admiral *(Limenitis populi)* on the leaves of an ash; newly emerged adult butterfly beside its cast off cocoon.

(Geometridae) and above all by tussock moths (Lymantridae)—the Black Arches Moth *(Lymantria monacha)* used to be the scourge of central European forests where it destroyed vast forests of spruce. The Noctuidae family includes adult moths of remarkable beauty but the tiger moths (Arctiidae) (Fig. 534) have many attractive larvae as well. The emperor moths (Saturniidae) (Fig. 536) are noted more for their patterning then for their coloration. The tropical Atlas Moth *(Attacus atlas)* of south-east Asia, however, has all three—beautiful wing shape, coloration and patterning.

540

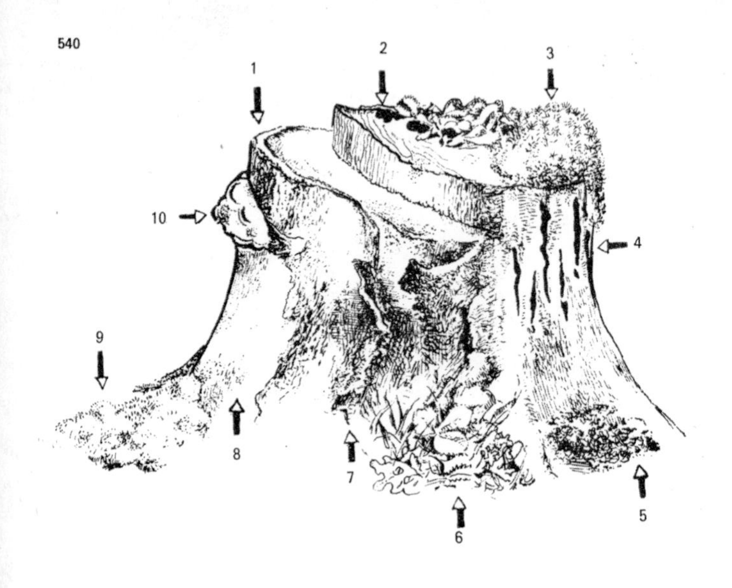

540 A beech stump inhabited by the larvae of various specialized families of two-winged flies (Diptera): 1 — under the close-fitting bark — Erinnidae and Asilidae, 2 — large holes filled with leaves — Psychodidae, 3 — under cushions of moss — Lycoriidae, Sciophylidae, 4 — in cracks — Syrphidae, 5 — in humus — Limnobiidae, 6 — in leaf litter — Cecidomyidae, Stratiomyiidae, Bibionidae, 7 — under the loose bark — Ceratopogonidae, 8 — in rooting wood — Tipulidae, 9 — under moss — Tipulidae, Therevidae, 10 — in polypores — Limnobiidae.

541

542

543

541 The Common Toad *(Bufo bufo).*

542 The European Treefrog *(Hyla arborea)* is equipped with suction discs for climbing trees.

543 The tropical toad *Bufo superciliaris* resembles a dry leaf casting a shadow.

544

544 The West African Hinged-back Tortoise *(Kinixys erosa)* lives in the undergrowth of the tropical forest.

545 The Nile Monitor *(Varanus niloticus)* in an African forest.

545

Noted for their narrow wings are the hawk moths (Sphingidae), some of which are pests that occasionally occur in vast numbers, such as the Pine Hawkmoth *(Sphinx pinastri).*

In the opinion of entomologists loveliest are the butterflies of the swallowtail family (Papilionidae). Those of the temperate regions are eye-catching but the greatest beauties are found in the forests of the tropics (Fig. 537 and 538). The family also includes the well-known, brightly coloured butterflies of the genus *Ornithoptera* of the islands of south-east Asia and Australia. The whites and yellows (Pieridae), blues, coppers and hairstreaks (Lycaenidae) and fritillaries and vanessid butterflies (Nymphalidae) are common inhabitants of forests throughout the world but only occasionally do they affect the normal development of the forest (Fig. 539).

The list of forest insects would not be complete if it did not include the two-winged flies (Diptera) (Fig. 540). Gall midges (Cecidomyidae) deform the leaves and flowers of forest plants and their larvae produce characteristic galls; mosquitoes (Culicidae) and horseflies (Tabanidae) suck the blood of animals and often transmit dangerous diseases; robber flies (Asilidae) and tachinid flies (Tachinidae) are parasitic on caterpillars or the larvae of other insects that may sometimes be so numerous as to defoliate trees. Flies of the Muscidae family, which includes the common housefly, lay their eggs in carrion and in the adult stage some species bite and suck blood, which makes them not only an annoyance but also possible transmitters of various diseases.

Amphibians

The forest is also the home of amphibians, even though these are creatures which require water for their development. However, they do not need a large river or lake—a small stream flowing under the canopy of the trees, and thus part of the forest, will suffice. The treefrogs of tropical forests even spend their tadpole stage in tiny pools of water in the junction between the branch and the leafstalk of epiphytes or tree branches, and the marsupial frogs *(Gastrotheca* sp.*)* do not require any water at all during the larval stage.

European forests are the home of the Spotted Salamander *(Salamandra salamandra)* which crawls slowly through the forest undergrowth early in the morning, feeding on worms, molluscs and insects. Numerous toads (Bufonidae) (Fig. 541) live in the forest undergrowth. One such is the tropical toad *Bufo superciliaris* which looks like a dry leaf casting a shadow (Fig. 543). Many species of frogs live in trees—these include members of the Hylidae (Fig. 542) and Dendrobatidae families and some Ranidae. The flying frog *Rhacophorus reinwardti* of the Rhacophoridae family can crawl but also has webbed toes which enable it to glide through the air.

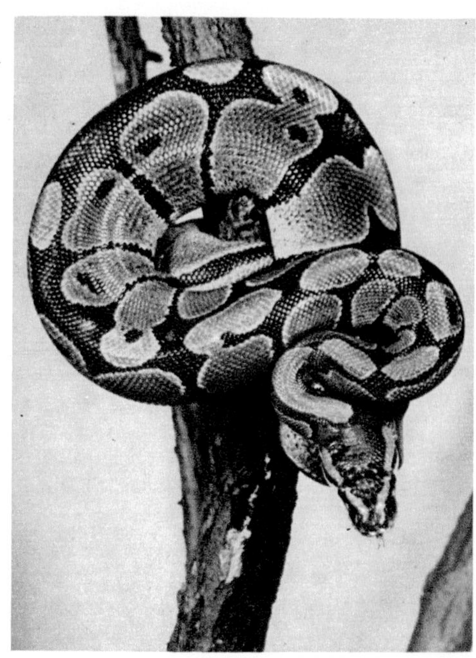

546

546 The Royal Python *(Python regius)* is native to the damp forests of Africa.

547 The Sand Lizard *(Lacerta agilis)* is a common European reptile.

547

548

Reptiles

Reptiles frequently found in the forest include the turtles (Testudines), lizards (Sauria) and snakes (Serpentes). Typical ground turtles of the forest undergrowth are those of the genus *Testudo* and the related genera *Homopus* and *Kinixys* (Fig. 544). Many lizards climb trees, like the geckos (Gekkonidae), iguanas (Iguanidae), chameleons (Chamaeleonidae) and monitors (Varanidae) (Fig. 545). High up in the treetops of the Malaysian rain forest lives the Flying Dragon *(Draco volans)* which has extended ribs on either side covered with membranous skin, which serve as wings for gliding from one tree to another. European forests are the home of certain smaller lizard species such as the Sand Lizard *(Lacerta agilis)* (Fig. 547).

Most forests are inhabited by snakes, even though in temperate regions they are mostly found in forest margins, thickets and clearings (Fig. 546). Snakes of tropical forests include forms that live mostly underground; that live mostly on the ground; and also many forms that inhabit trees. Subterranean snakes, such as the African species of the genus *Typhlops,* have a smooth body resembling a large earthworm. The two ends are practically identical and it is difficult to distinguish the head with its blind eyes. These snakes feed mostly on small soil insects and only occasionally do they come to the surface. Other forest snakes, like the Gaboon Viper *(Bitis gabonica)* (Fig. 548), pythons and cobras, hunt their prey—small rodents, reptiles or amphibians—mostly on the ground. Typical snakes of the forest ecosystem are the tree snakes that live in the treetops and on trunks. One European example is the Aesculapian Snake *(Elaphe longissima),* while African forests are the home of the Boomslang *(Dispholidus typus).* There are far fewer snakes in the undergrowth of tropical forests than high up in the treetops which is why foresters generally come across them either at the edge of the forest or in clearings, particularly if these were used for the cultivation of agricultural crops attracting rodents and frogs. The greatest diversity of snake spe-

549

548 The Gaboon Viper *(Bitis gabonica)* is practically invisible on the floor of the virgin forest even though it appears to be strikingly coloured.

549 Nest of the Black Stork *(Ciconia nigra).*

550 A Goshawk *(Accipiter gentilis)* with its victim.

551 A young Goshawk *(Accipiter gentilis)*.

550

551

cies is found to exist in tropical rainforests interspersed with cocoa plantations.

All snakes climb well but the best tree climbers are certain boas and pythons (Boidae), many Colubridae and some Elapidae. Snakes include among their number a 'flying' snake— *Chrysopelea ornata*— which does not actually fly but is capable of gliding from tree to tree by forming a U-shaped groove on its ventral side which aids it in its flight. It inhabits the forests of Ceylon, south India and south-east Asia.

Ground as well as tree species all have very intricate protective coloration not only as regards combination of colours but also patterning. The Gaboon viper has an unusual geometric pattern with complex distribution of brown, blue, green, reddish-brown, yellow and black. On the dorsal side there is a pronounced pale interrupted band which in the undergrowth blends with the surroundings to look like twigs scattered in the forest litter. The Gaboon viper is dangerous in that it is inconspicuous and is therefore easily overlooked

552

552 The Ring-necked Pheasant *(Phasianus colchicus)* is a bird of open forests and hedgerows.

553

and trodden on. The bite of such a snake, whose fangs are up to five centimetres long and whose poison has a damaging effect on the vascular as well as the nervous system, usually ends in death. There are many aggressive snakes in forests throughout the world but they usually attack only smaller prey such as rodents, birds or amphibians. Man is in danger of being bitten only if the snake is cornered by him, either by accident or intention.

Birds

That part of the forest ecosystem which spreads high above the ground demands of its inhabitants either that they be able to climb well or fly. The best fliers are the birds (Aves) and of the mammals, the bats (Chiroptera). The flying ability of insects should not be underestimated, but with regard to the more robust structure of vertebrates the problem of flight in the case of birds and bats has been solved with great success. These animals

554

555

553 The Hobby *(Falco subbuteo)* keeps to the treetops.

554 The Eurasian Eagle Owl *(Bubo bubo)* was forced by civilization to take refuge on inaccessible cliffs.

555 The Little Owl *(Athene noctua)* feeds mostly on small mammals.

556

556 A Great Spotted Woodpecker *(Dendrocopus major)* in its typical position on a tree trunk.

range over a large territory and the forest is generally only a temporary home, serving either as a nesting ground or as a source of food.

Of the wading birds—herons and their allies (Ciconiiformes)—one forest inhabitant is the Black Stork *(Ciconia nigra)* (Fig. 549) for it nests in trees, though it seeks its food on water, sometimes some distance away. Another that nests in the forest, in treetops near water, is the Night Heron *(Nycticorax nycticorax)*. Of the waterfowl (Anseriformes) the Common Goldeneye *(Bucephala clangula)* makes its nest in cavities in old trees. Birds of prey are an important group in the life of the forest. Where beasts of prey are absent they form the top level of the food pyramid of the forest ecosystem (Fig. 550, 551 and 553).

Important game birds in Europe and North America are the fowl-like birds (Galliformes) which include the Black Grouse *(Lyrurus tetrix)*, Ring-necked Pheasant *(Phasianus colchicus)* (Fig. 552), Wild Turkey *(Meleagris gallopavo)* and many other related species. Of the waders (Charadriiformes), associated with life in the forest is the Woodcock *(Scolopax rusticola)*, which is also a game bird. Pigeons and their allies (Columbiformes) are mostly tree and forest birds. Many species of pigeons feed on the fruits and seeds of forest plants. These also make up the diet of parrots, which are found in tropical forests. Their strong beaks are designed to crack hard seeds.

Many forest animals are active at night so during the course of evolution nature has produced nocturnal hunters. Owls (Fig. 554 and 555) are equipped for this purpose with a remarkably keen sense of sight and hearing. Well-known for their characteristic call and method of rearing their young are many species of cuckoos (Cuculiformes) (Fig. 558). Their cry is always evocative of forests in early summer. Of the swifts and hummingbirds (Apodiformes) it is the hummingbirds that are important to forests. They fly like insects round tropical blooms feeding on nectar as well as on the insects crawling there. Birds related to the kingfishers also include typical forest inhabitants like the motmots (Momotidae) of the tropics, the hoopoes (Upupidae) (Fig. 557) and the hornbills (Bucerotidae). Hornbills are noted for their strong but light beaks and unusual nesting habits. As a protection from snakes, they nest in tree cavities, the female being sealed in by the male, who provides his mate and offspring with food until the nestlings are fully grown. Food is supplied

Dam built by the Canadian Beaver *(Castor canadensis)* Vermont, U.S.A.

Mountain mist forest on the Pacific coast near Acapulco (Mexico).

557

558

through a small slit left in the plastered-up hole for that purpose.

Typical inhabitants of forests the world over are the woodpeckers and their allies. These include the woodpeckers proper (Picidae) (Fig. 493 and 556), the honeyguides (Indicatoridae), toucans (Ramphastidae) and other tropical families. The largest order of the avian realm are the perching birds. Included in this group are the ovenbirds (Furnariidae), tyrant flycatchers (Tyrannidae) and bellbirds (Cotingidae)—all mostly tropical families. Skillful at catching insects on the wing, as their name implies, are the Old World flycatchers (Muscicapidae) (Fig. 560). The Old World warblers (Sylviidae) are to be found in practically all forested areas—the willow warbler is a typical example (Fig. 559). Closely related to the warblers are the thrushes (Turdidae), one of which is the European Blackbird *(Turdus merula)* which, though originally a forest inhabitant, has become one of the commonest birds of

557 A Hoopoe *(Upupa epops)* with its prey in front of its nesting cavity.

558 A young Cuckoo *(Cuculus canorus)* making its well-known cry.

559

560

559 A Willow Warbler *(Phylloscopus trochilus)* by the nest.

560 A Collared Flycatcher *(Ficedula albicollis)* beside its nesting cavity.

561 A European Blackbird *(Turdus merula)* with its young.

562 A Wren *(Troglodytes troglodytes)* feeding its young.

563 A male Red-backed Shrike *(Lanius collurio)*.

561

562

cultivated areas (Fig. 561). Only one member of the American family of wrens (Troglodytidae) (Fig. 562), which feeds on insects, is found in European forests. Shrikes (Laniidae) (Fig. 563), consume insects and small vertebrates. Flocks of tits are continually seeking food on the branches of trees, and even in cultivated forests they continue to be the best means of controlling insect populations (Fig. 564, 565 and 566). Though it feeds on insects in summer, in the winter months the Waxwing *(Bombycilla garrulus)* (Fig. 567) gathers forest fruits, mostly berries that remain on the branches of woody plants in winter.

The role of pollinators played by hummingbirds is taken on in the African and Asian tropics by the sunbirds (Nectaridae). Common throughout America are the wood warblers (Parulidae), and in the New World tropics the tanagers (Tanageridae). Important birds in the temperate forests of Eurasia and North America are the New World seedeaters (Fringillidae), which include the Bullfinch *(Pyrrhula pyrrhula)* (Fig. 569), the Common or Red

563

564

565

564 The Coal Tit *(Parus ater)* consumes a large quantity of food.

565 The Great Tit *(Parus major)*.

Crossbill *(Loxia curvirostra),* and the well-known Chaffinch *(Fringilla coelebs).* Practically the whole of Europe is home to the Common Starling *(Sturnus vulgaris)* (Fig. 568). Originally the starling nested in tree cavities and foraged for insects in clearings but as such natural cavities became increasingly fewer it adapted itself to the conditions and now nests in cracks in walls, under eaves, and frequently also in man-made nestboxes.

The forest is also the home of numerous members of the crow family (Corvidae). No European forester could imagine a forest without the Nutcracker *(Nucifraga caryocatactes)* though it is very rare in Britain. Much better known is the Jay *(Garrulus glandarius)* and the Carrion and Hooded Crows *(Corvus corone).*

Mammals

The world's forests also shelter many diverse groups of mammals. The marsupials or pouched mammals (Marsupialia) are found only in the forests of America and Australia. Endowed with remarkable vitality is the Opossum *(Didelphis virginiana),* which is omnivo-

566

566 Eggs in the nest of the Great Tit *(Parus major).*

567 The Bohemian Waxwing *(Bombycilla garrulus)* feeds on various fruits.

567

568

569

rous, feeding chiefly on rodents but also on plant food. In Australia the marsupials occupy many important niches in the forest ecosystem. Typical tree inhabitants are the cuscuses (Phalangeridae) as are some kangaroos, such as *Dendrolagus lumholtzi,* and of course the well-known koala.

Insectivores are adapted to life on the ground and below the surface of the soil. They feed chiefly on arthropods, worms, molluscs, and occasionally on the flesh of larger animals, though plant food also forms part of their diet. Common in Old World forests is the European Hedgehog *(Erinaceus europaeus)* (Fig. 571), the only spiny mammal in central Europe. More numerous are shrews and moles (Fig. 570) which consume large quantities of food daily. The floor of America's forests is riddled with the tunnels and covered with the mounds of the American Garden Mole *(Scalopus aquaticus),* and that of European forests, mainly forest clearings, by those of the European Mole *(Talpa europaea)* (Fig. 572).

Fitted for nocturnal life, particularly in the upper layers of the forest, are the bats (order Chiroptera). The fruit bats of the suborder Megachiroptera feed chiefly on plant food. Some lap the nectar of flowers thus helping to pollinate the trees that produce these blossoms. Bats of the suborder Microchiroptera (Fig. 574) feed primarily on insects. Those that do not leave the confines of the forest sleep in hollow trees during the daytime.

South American forests are the home of 'toothless' mammals of the order Edentata, namely five species of sloths and two species of tree-dwelling anteaters.

Forest rodents (Rodentia) are herbivores and are therefore adapted to a diet of plant food. They live underground, on the surface, and high up in the treetops. Excellent at climbing are tree squirrels (Sciuridae), found in both European and American forests. The Grey Squirrel *(Sciurus carolinensis)* (Fig. 577), a native of North America, was introduced to Britain in the 1870s and has become a serious pest, damaging hardwood trees, particularly beech and sycamore, by extensive and apparently pointless bark peeling. Flying squirrels (Petauristiae), which are related to the tree squirrels, have broad membranes between the front and hind legs on either side which are expanded when the squirrel leaps, thus helping it to glide through the air as it travels from one tree to another. Several species of flying lemurs are similarly equipped.

Of the rodents the Canadian beaver plays a dynamic role in the forest, as did the European beaver at one time. The long-tailed rodents of the Muridae family have the greatest number of species. Among them are the many fieldmice of the genus *Apodemus* (Fig. 575) and the large rats of the tropics, as well as voles such as the Bank Vole *(Clethrionomys glareolus)* which even occasionally crawls on the bottom branches of trees. Dormice (Muscardinidae) are widely distributed but only in Eurasia. Porcupines (Erethizontidae) are commonly found on trees in the forested regions of North America. The Canadian Porcupine *(Erethizon dorsatum)* (Fig. 576) feeds on the bark of forest trees.

Beasts of prey play an important role in maintaining the balance of the forest ecosystem. They have very keen sense organs, are extremely agile and superbly equipped for

570

568 A Common Starling *(Sturnus vulgaris)* beside its nesting cavity.

569 A Bullfinch *(Pyrrhula pyrrhula*

570 An Alpine Shrew *(Sorex alpinus).*

571

572

573

571 The European Hedgehog *(Erinaceus europaeus)* eats a great variety of foods.

572 The European Mole *(Talpa europaea)* leaves its underground passages only occasionally.

573 Edible or Fat Dormouse *(Glis glis).*

574 The Noctule Bat *(Nyctalus noctula)* is found in most parts of Europe and Asia.

574

capturing live animals. All forests throughout the world contain at least one or more flesh-eaters of the cat family (Felidae) (Fig. 578). Largest of these is the tiger but the most skilled and most versatile hunters are the leopard or panther, the jaguar and the puma or mountain lion. Those that make their home in Europe's forests are the Wild Cat *(Felis sylvestris)* and the lynx. Other beasts of prey that inhabit forests are members of the dog family (Canidae). Best known is the Wolf, which is found in the northern hemisphere (Fig. 579). The weasel family (Fig. 580, 581) contains mostly small beasts of prey, except for the glutton or wolverine, and the Badger. The Badger *(Meles meles)* is a common woodland animal in Britain, excavating labyrinthine setts (burrows) beneath the trees from which it emerges only at night-time to amble along well worn tracks in search of worms, grubs, carrion and any smaller creature too young or too inactive to avoid its lumbering progress (Fig. 583).

575

575 Long-tailed Fieldmouse *(Apodemus sylvaticus)*.

576 Canadian Porcupine *(Erethizon dorsatum)* in a forest clearing in Alaska.

576

577

7 The Grey Squirrel *(Sciurus carolinensis)* was originally a native of North America, but in the forties of the last century it was introduced to Great Britain. It became a serious pest because its numbers reached astounding proportions and its vast populations nibble the bark of the broadleaved trees, chiefly beeches and sycamores.

578

579

Elephants (order Proboscidea), represented in Africa by the African Elephant and in Asia by the Indian Elephant, can have an important influence on the composition of the forest. Of the odd-toed ungulates (Perissodactyla) those that occur in forests are the rhinoceroses (Rhinocerotidae) and tapirs (Tapiridae). Far more widespread in forests are the even-toed ungulates (Artiodactyla). Members of the pig family (Suidae) are always conspicuous elements in the forest undergrowth (Fig. 582) and members of the deer family (Cervidae) are to be found in most forests, unless they have been exterminated by man. Well known are the red deer *(Cervus elaphus)* (Fig. 587), Sika deer *(Cervus nippon)*, fallow deer *(Dama dama)*, roedeer *(Capreolus capreolus)* (Fig. 586), white-tailed deer *(Odocoileus virginianus)*, European elk *(Alces alces)* and the reindeer *(Rangifer tarandus)*, which moves far into the forest belt in winter. A rare even-toed ungulate of the forest, however, is the okapi, which is found only in the rain forest of the Zaire River basin.

Forests in many parts of the world were inhabited, or still are, by herds of hollow-horned ungulates (Bovidae). The European bison was a typical forest animal but is now on the verge of extinction. The American prairies were the

578 The Puma or Mountain Lion *(Puma concolor)* is found throughout North and South America.

579 The Red Fox *(Vulpes vulpes)* has a vast area of distribution.

principal range of the North American bison but neighbouring forests were invaded by these beasts on their migratory journeys south (Fig. 585). The virgin forests of Africa are the home of many species of antelopes, such as bushbucks *(Tragelaphus scriptus)* and forest duikers.

As typical tree-dwelling animals, primates are characteristic inhabitants of the forest. They are divided into two suborders: the Prosimii (Fig. 590), which include the lemurs, bushbabies, tree shrews and tarsiers; and the Anthropoidea, which include the monkeys, baboons and apes. Prosimii are nocturnal creatures with excellent sight and sense of smell. The tree shrews (Tuapidae) of south-east Asia are carnivores or omnivores. The lemurs (Lemuridae) of Madagascar and tarsiers (Tarsiidae) of Malaysia are apparently omnivores. Chief representatives of the Anthropoidea on the American continent are the New World monkeys of the family Cebidae (Fig. 589), which travel in troops in the tree-tops feeding mainly on juicy twigs and fruits. The Old World anthropoids are divided into three large families: the Old World monkeys (Cercopithecidae), the gibbons (Hylobatidae) and the anthropoid apes (Pongidae). These are mostly herbivores or even fruit-eaters. Where-ever large troops dwell they affect the forest ecosystem both directly by what they consume and indirectly by tolerating only certain animals on their territory.

580

580 A Pine Marten *(Martes martes)*.

581 The Stoat *(Mustela erminea)* is a typical representative of the weasel family.

581

582

583

582 A group of Wild Boars *(Sus scrofa)* foraging for food under a fruiting oak.

583 The Badger *(Meles meles)* builds a complex system of underground chambers and passages under the roots of trees which it leaves only at night when it sets out for hunt along its well-trodden paths. It feeds on worms, grubs, carrion and various small animals, especially young and diseased animals.

584 Running Reindeer *(Rangifer tarandus)* at the polar timberline in Alaska.

585 North American Bison *(Bison bison)* in the forests of Alaska.

584

585

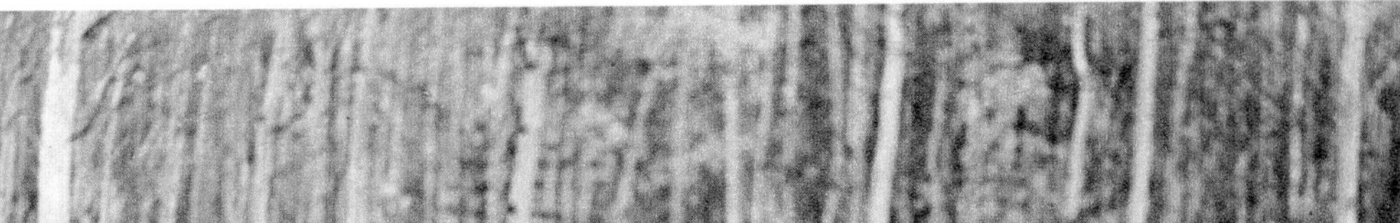

586

586 Roedeer *(Capreolus capreolus)* in velvet.

587 Herd of Red Deer *(Cervus elaphus).*

588 The Spider Monkey *(Ateles geoffroyi)* makes good use of its prehensile tail when moving about in the treetops.

589 The Brown Capuchin Monkey *(Cebus apella)* of South America.

590 The Lemur *(Lemur mongoz)* is well equipped for climbing trees and seeing in the dark.

587

588

589

590

591

591 Log dump consisting mostly of telegraph poles.

THE RENEWABLE SOURCE

Forest management

Through countless years man exploited the forest, felling the trees arbitrarily and paying no attention to the fact that the tall forests around developing areas were disappearing and being replaced by worthless thickets, grassy clearings, dry meadows or rocky banks. The forestless landscape round farms and villages gradually increased in size and the once small enclaves surrounded by forest joined to form continuous agricultural land. Gradually man had to go further for fuel, timber, medicinal drugs, clear springs, wild game and recreation. Nowadays transport makes it possible to obtain forest products from neighbouring forested regions, from other countries or even from overseas, but at some future time such methods may become commercially unprofitable and prohibitive. At this point the citizens of agricultural or agricultural-industrial countries must solve the problem of what to do next.

The solution is planned logging and organized silviculture—in other words sensible forestry (Fig. 595 and 596). Forests should be exploited only in such a way as to ensure the continual regeneration of useful plants and animals without reducing the fertility of the soil or causing permanent deterioration in the climate near the ground; a way that ensures the constant or more rapid height and diameter increment of marketable trees, maintains the numbers of wild game at the desired level, and ensures that the forest receives the greatest possible amount of rainfall, uniformly passing it on, unspoiled, to springs and water courses. Forests cannot be a no-man's-land, a wilderness without a responsible caretaker. Forests must be managed by experienced and trained foresters. All densely populated and developed nations have some form of forestry. However, in large parts of the world man's actions in the forests continue to be haphazard and without regard to the results.

The first requirement of rational management in the forest is order. The boundaries of the area for which the forester is responsible must be clearly defined and he must have a long-term forest management plan. The vastness and diversity of forest reserves and the longevity of forest trees makes organization in time and space a far from simple matter, which is why a specialized branch of forestry known as forest management has come into existence. Its basic principles are the same the world over, even though in tropical regions it is necessary to modify the methods that have been previously tried and tested in European and North American forests.

A forest reserve, then, must be precisely surveyed, drawn on a map and demarcated in the field (Fig. 592 and 593) so that the work of the forester is not at cross-purposes with the needs of his neighbour, and so that it is possible to rely upon a certain production area and a certain stock of trees. This clear and permanent demarcation is particularly important where forest areas border upon pastures and arable land. It was not until the eighteenth and nineteenth centuries that order was established in the forests of central Europe and the first forest ordinances required that the boundaries of forest reserves be marked by stone boundary pillars and by ditches or cleared rides round the periphery. In many parts of the world it is not easy to maintain permanent boundaries. African foresters, for instance, employ hundreds of forest workers merely to keep the rides round the forest reserves cleared. Villagers are free to pursue their own interests in the surrounding jungle so if the boundaries were not clearly marked, parts of the state forest could easily be damaged. In the dry Australian bush the ride marking a boundary must be very broad because of the risk of fire which can easily leap over the intervening space.

For the forester's needs a boundary demarcated only in the field is not enough. He must have an exact plan or map of his district (Fig. 594 and 595) in order to record changes, and also to know where trees earmarked for logging or treatment are located. Forest surveying is a difficult profession because forests are by nature a mass of tangled vegetation, difficult of access. Thanks to old traditions, however, and also nowadays to sophisticated surveying instruments and aerial photography, foresters the world over are provided with good, detailed maps showing the composition of the forest and features of the forest environment. These maps serve as the basis for forest management plans.

592

593 A stone boundary pillar demarcating a forest reserve.

592 A ride dividing a forest complex into smaller sections.

594 Measuring the area and boundary of a forest with a simple surveyor's compass.

Forestry cannot be governed and carried out from a remote office. Those who are responsible must live in the forest. The architecture of foresters' houses, forest district offices and large forest enterprises is always influenced by the environment and available material. The combination of wood, stone and brick (Fig. 598) is a very good one in damp cool regions. Foresters in the tropics are faced with the problem of troublesome insects (Fig. 599). Wooden foundations and walls are easily destroyed by termites and concrete must therefore be used where necessary. Mosquitoes are an unpleasant nuisance and thus all windows and porches must be covered with nettling.

Regeneration

The forester's chief responsibility is the continual regeneration of the forest. Felled trees or stands damaged by natural catastrophes must be continually replaced by new ones. The regeneration of the forest is governed by the method used for logging, and also by the aims of the forester, which as a rule are not identical with the natural development of the forest. If only single trees or small groups are felled, then natural regeneration can take its course with successful results. The seeds of mature trees in the immediate vicinity fall in the open space and germinate, or saplings growing there start to spread (Fig. 600 and 601). This is easy in the case of trees and shrubs that tolerate shade during their youth, for they can make the most of the shaded microclimate, but trees and shrubs that require sunlight for their development require larger natural spaces or man made clearings.

Botanists and ecologists once thought that the forest regenerated naturally only in small gaps and that small-scale regeneration alone would guarantee the health of cultivated forests. In actual fact, long before the advent of man, numerous natural catastrophes such as gales, floods and fires created huge clearings where the forest regenerated spontaneously. Thus even in natural forests there are instances of regeneration in both small and large areas, and the silviculturist could follow the same example with an eye to the desired timber yield or other functions of the forest.

On large deforested areas the forester generally resorts to artificial regeneration, which involves the deliberate planting of a selected

593

594

595

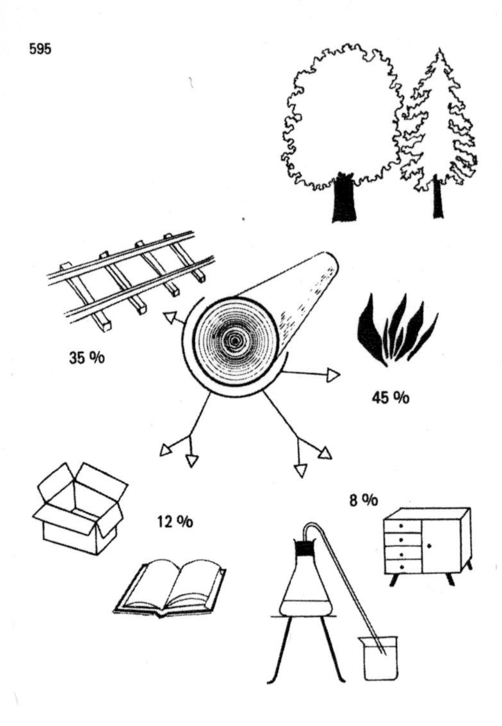

species or mixture of woody plants (Fig. 603). With suitable border cutting, however, even large clearings in time contain much new growth from natural seeding (Fig. 604) and it is possible to combine artificial regeneration with natural regeneration. The former provides the opportunity for the forester to shape the future development of the forest according to the needs of mankind by filling the cleared spaces with mixed species of top quality plants, suitable for the given environment as well as for the demands of commerce.

For this purpose the forester may use seedlings and saplings taken from areas with rich natural regeneration. This procedure is followed by foresters of tropical regions, where fruiting trees with rich crops of seeds are unevenly distributed throughout the area. Lifting seedlings from beneath self-sown trees, however, is very uneconomical and the plants are usually of inferior quality because their root systems are damaged when they are pulled from the forest soil. However, with the aid of modern equipment it is possible to lift large saplings and trees—a procedure widely used nowadays in landscaping housing developments and laying out gardens.

Most planting material for artificial regeneration comes from forest nurseries. Modern techniques result in nurseries producing plants that root well and begin rapid growth within a short time of transplanting. The most important feature of good plants is a well-branched root system that is not too deep.

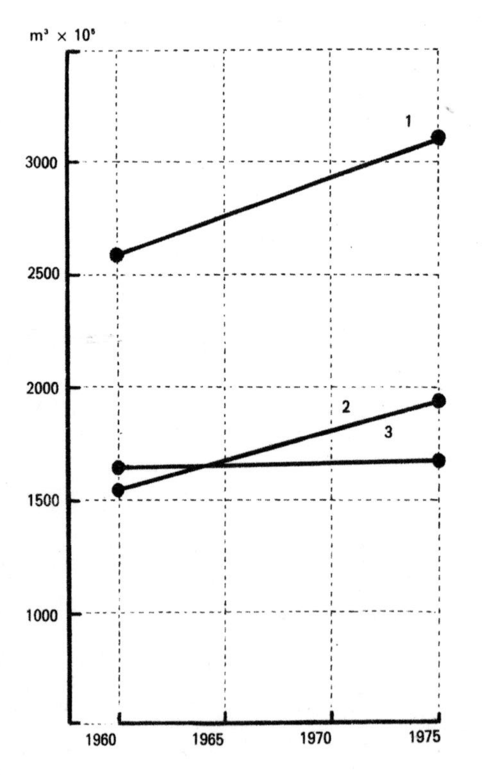

596

595 Percentage consumption of wood by various sections of the economy.

596 Trend in wood consumption in the years 1960 to 1975: 1—overall consumption of wood, 2—consumption of timber, 3—consumption of fuelwood.

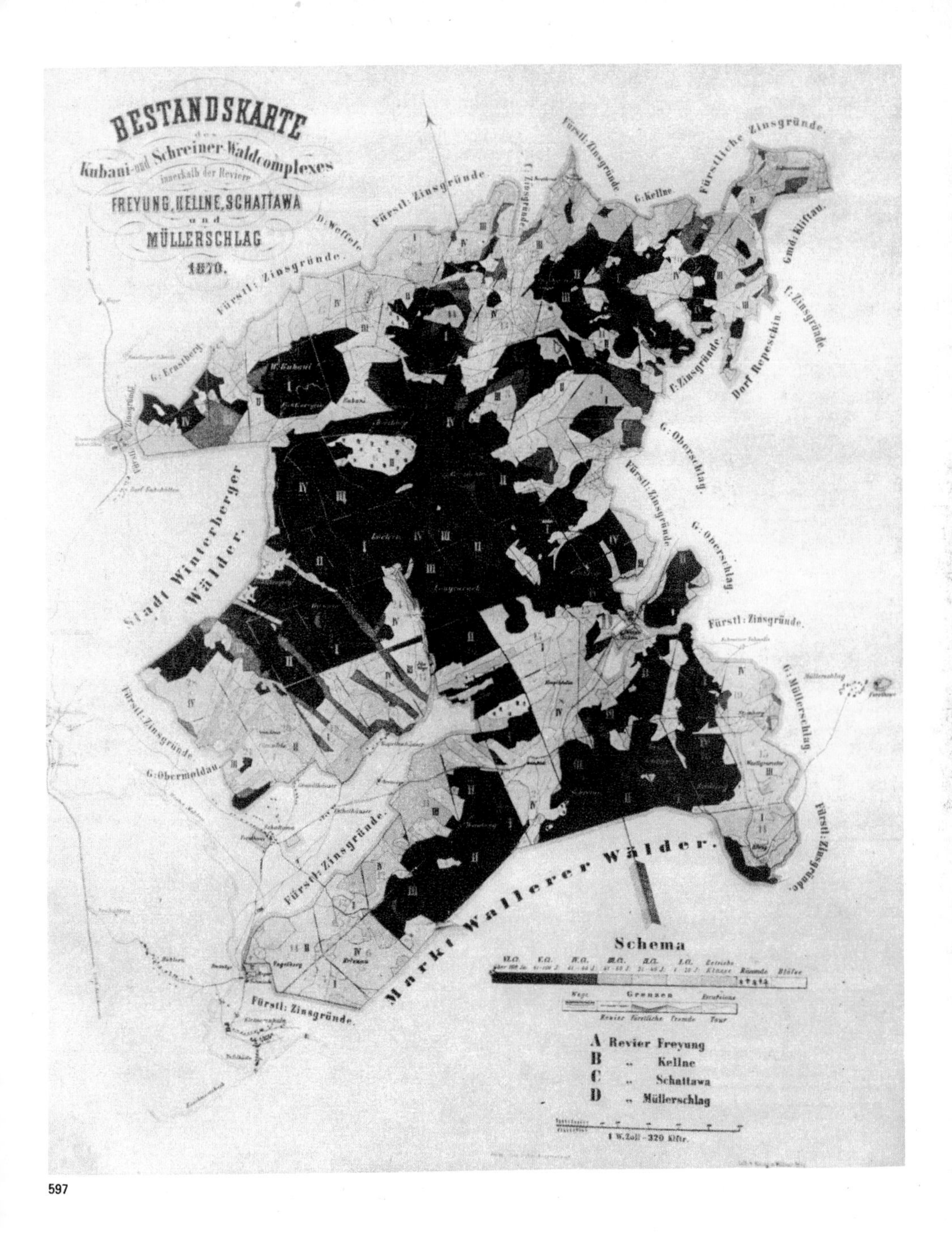

597

597 This forestry map of 1870 served as the silviculturist's basis for his organization in time and space.

598

598 A rustic house is a fitting dwelling for foresters.

599 A forester's house in the tropical rainforest must be built to withstand termites and be fitted with protection against mosquitoes; Ghana.

599

A conspicuous forest reserve boundary of conifers in the Australian bush.

An old forestry map with stands of various ages marked in different colours.

Exotic conifers *(Picea sitchensis, Larix* spp.*)* are planted in pasture lands covered with Bracken *(Pteridium aquilinum);* Queen Elizabeth Forest Park, Scotland.

An extraction cableway in operation in the mountains of Czechoslovakia.

600 Natural regeneration of stands in the mountains of Japan.

601 Natural regeneration of spruce in the mountains of central Europe.

600

601

602

602 Enrichment planting in a tropical rainforest; the planted trees belong to the species *Khaya ivorensis* and are grown for reasons of commerce.

This may be achieved by undercutting the plants in the nursery bed at a certain period of growth by moving a knife through the ground about ten centimetres below and parallel to the surface thus severing the tap root and promoting the growth of lateral roots. Great success has also been achieved with seedlings grown in peat pots. In such a pot, roots form round the margin of the pot because peat promotes root branching, and seedlings root well when put out in their permanent site.

The success of artificial regeneration depends in great measure on how the seedlings are planted. In loose and grass-free soil the roots may simply be inserted in a hole or a gash made with a hoe. Where the soil is covered with forest weeds, however, the ground must be hoed and loosened over a fairly large area. Sometimes mounds are prepared to receive the seedlings. For practical purposes, mainly for checking the number of seedlings planted out on a given area, a specific pattern with regular spacing is used for planting. Regular spacing also facilitates tending.

In the temperate regions of Europe and North America the best time to plant out seedlings is just before the plants begin to put out leaves in spring, while in subtropical forests with their alternating rainy and dry seasons the best time is shortly before the first rains. In

603 Hoe-planting—the method used for planting a large number of trees in Europe.

604 Border cutting in a mountain beech forest promoted the natural regeneration of scattered spruce.

603

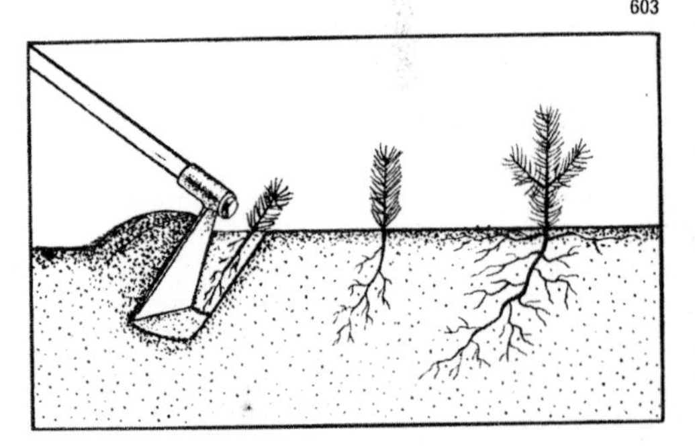

604

the evergreen rainforest it is hard to choose a time, for individual species of merchantable trees seem to have their own inner rhythms that have little to do with the human calendar.

Artificial regeneration is just as important in tropical forests as it is in temperate regions. The presence of luxuriant growth suggests that there are plenty of natural seedlings to provide for the continuity of the line. From the forester's viewpoint, however, it is necessary to ensure the artificial regeneration of merchantable species which rarely regenerate in sufficient numbers by natural means. This is done by so-called enrichment planting (Fig. 602). Parallel lines are cut through the forest and in these are planted robust saplings of selected species. Saplings up to three metres tall are used so that they are easily recognised when routine weeding is carried out within the forest.

The first years in the life of a tree are always critical ones and therefore special measures must be taken when planting to limit the unfavourable influences of the environment (Fig. 605). In forest reserves containing a great number of red deer and roe deer all areas with newly planted saplings must be protected by means of temporary fencing (Fig. 606) or by applying animal repellents to the buds and twigs of the young plants.

The greatest danger for young plants in gaps and clearings is the competition of herbs, shrubs, lianas and commercially worthless trees. Stronger light, and nutrients released by the decomposing humus promote the growth of weeds as well as of planted trees. In European forests these include Groundsel *(Senecio vulgaris)*, Rosebay Willowherb *(Chamaenerion angustifolium)*, Raspberry and Blackberry bushes *(Rubus idaeus* and *R. fruticosus)* and Bush grass *(Calamagrostis epigeios)*, later augmented by trees and shrubs such as elder, birch, goat willow, aspen and rowan. In tropical forests they include mostly fast-growing lianas and light-demanding trees. One of the difficult tasks required of the forester is to aid the new plants in their competitive struggle against their aggressive neighbours. The sickle, scythe and machete are still the most effective tools but lack of labour in modern countries has brought about the use of specific poisons—herbicides and arboricides—which are not without their biologically harmful aspects.

Plants grown in forest nurseries are also used for the afforestation of devastated areas. On bare sands (Fig. 608), steep banks (Fig. 607), in barren limestone mountains (Fig. 610) and on weed-infested pastureland, the forester must expend particular effort in order

605

606

607

608

605 Planting on raised mounds makes it possible to regenerate a spruce forest.

606 A clearing in a coniferous forest planted with beech *(Fagus sylvatica)* must be protected with fencing against browsing by wild animals.

607 Reafforestation on steep devastated banks requires the use of wicker fences.

608 Reafforestation of barren sands.

that the young trees may survive the adverse influences of drought, insufficient nutrients, erosion by wind and water, the heat of fires, overpopulations of insects and the competition of weeds. The return of forest to places where it has been irresponsibly destroyed takes effort, time and money, but here, too, science, technology and man's ingenuity have met with success. In some places devastated areas are again covered with the greenery of young forests. The soil is gradually being enriched with humus, nutrients and water and the extremes of climate near the ground are being lessened by the overhead protection of the canopy.

Naturally regenerated or cultivated plantations must be tended by the forester at all stages of growth. In the course of their development he can control the tree density and the composition of species and also influence the quality of the growth forms. It is likewise necessary to keep an eye on the health of

609 A naturally regenerated young beech wood prior to clearing.

610 Only with the aid of man-made terraces was it possible for the forest to return to devastated banks in the Sierra Nevada Mountains of Spain.

610

young stands because harsh climatic conditions, damage by wild game, fungus diseases and the depredations of insects are forever claiming new victims.

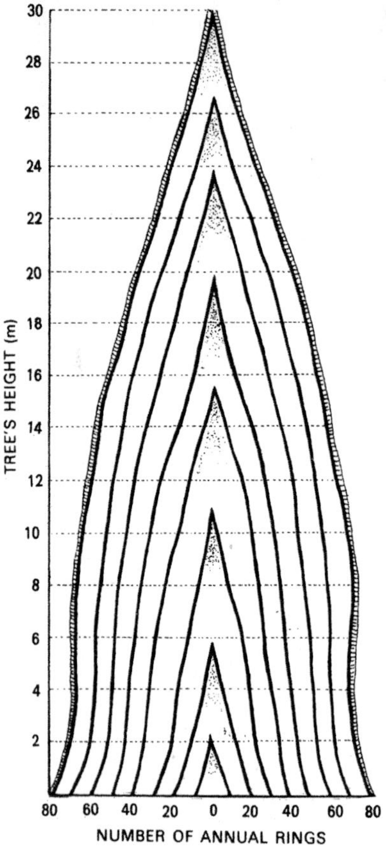

Congested areas must be thinned (Fig. 609), and thinning by the forester speeds up the process of self-thinning which slowly but continually takes place in every forest. Sooner or later certain strong individuals become dominant and neighbouring trees overpowered by their shade and root competition begin to dry up. This is accompanied by marked losses in growth which can be kept within bounds only by timely thinning and influencing the development of the young growth in favour of the strong, quality trees. In young mixed forests it is already necessary at this stage to promote the growth of merchantable species.

In older woodlands the work of the arborist who tends the stands becomes influenced by the interests of the consumer. Tending becomes aimed at promoting the growth of quality trees (Fig. 611). This may be achieved either by negative selection, which is by continual seeking out and felling of diseased and crooked trees, or by positive selection, which involves looking for promising trees and opening up the crown, as well as clearing the ground around the roots to provide the greatest amount of space. High thinning, which takes place in the main part of the canopy, is a far more effective method of influencing the development of the forest than low thinning (the felling of smaller trees).

The method of thinning is also determined by the required yield. Ideally, trees should be thinned little and often. In practice economic pressures force the present-day forester to carry out more radical thinning at lengthier in-

611 Longitudinal section of a tree trunk with marked layers of ten annual rings; when the trunk is cut into two-metre sections it is possible to determine the height of the tree at a certain age, as well as the volume increment.

612 This mechanism automatically climbs up the trunk, lopping off branches.

613

614

tervals. Primary interest in a certain species of tree may override other principles. For example the vast demand for the wood of the yellow birch forces North American silviculturists to thin stands so as to promote the growth of this tree even though other useful and merchantable hardwoods are suppressed as a result.

The silviculturist is also concerned with growing straight, high quality boles. Until recently the only existing method of removing thick, persistent branches on the lower part of the trunk was to cut them off by hand, while the only way to get to branches higher up was to use a ladder. Nowadays use is made of a mechanism that can climb up the trunk like a squirrel and lop off the branches (Fig. 612).

Another of the forester's duties is to protect

613 A young beech peeled by red deer.

614 A young beech spot-painted with a mixture of wood tar, clay and sand to repel red deer.

the forest against damage and mutilation. Great losses are incurred as a result of red deer peeling the bark from young trees. The best method of prevention is to feed the wild game and control the size of the population. Sometimes endangered trees are coated with repellents (Fig. 614). Foresters in the large reserves of the tropics are faced with similar problems. For example the African elephant is capable of ruthlessly peeling the bark from trees and even uprooting them to get at the few ripe fruits in the crown.

Seed production

Artificial regeneration requires forest seedlings and these are grown in nurseries from seed. Formerly man was content to collect fruits and seeds lying on the ground under fruiting trees. To obtain a sufficient quantity of seed from tall, quality trees, however, was not easy and the practice using seed gathered from low and crooked trees proved a costly mistake in European forests. Nowadays seed is gathered from certified seed stands or even from single selected trees (Fig. 615). Grown from these seeds are plantations that will one day be seed orchards producing suitable seeds for forest regeneration.

The ripe cones of conifers or the fruits of angiospermous hardwoods must be harvested high up in the treetops, often thirty to forty metres above the ground. Specially trained seed collectors (Fig. 616) climb to such heights with the aid of climbing irons and safety belts. Their counterparts in the tropics, who have always climbed trees in this way, are the natives who climb slender palm trees with the aid of two rings woven of lianas. One of these is stretched between the man's feet and the other encircles his waist and the treetrunk, facilitating upward movements. There is no doubt that high quality seed merits the same artistic performance as the palm wine or nuts

615

616

617

615 Larches *(Larix decidua)* selected for quality seed production and grafts.

616 A cone collector climbing a tall pine.

617 A nursery in a tropical rainforest.

618

618 Cultivation of forest seedlings in a greenhouse.

619 Work in a large nursery can be mechanized.

that forest dwellers climb the heights to reach.

The height of the trees is not the only obstacle to obtaining a sufficient quantity of seed. In the case of species which are widely dispersed through the forest, it is never easy to find enough fruiting trees. A common difficulty in tropical forests is in determining the best time for fruit or seed collection. The seeds of most tropical trees have a very brief period of viability. When to this is added the fact that in some species the fruits of individual trees ripen at different times it is easy to understand the difficult situation of the forester who needs a large amount of seed for the purpose of artificial regeneration.

Very occasionally in the case of tropical species and more frequently in the case of trees of temperate forests (yew, ash and lime) the seeds do not germinate at all in the first year after the fruit ripens. They require a long time to mature physiologically and for moisture to soak through the external cover from the outside.

The seeds of forest trees of temperate regions remain viable for several years. For example, the seeds of Norway spruce have a high percentage of viability for a period of four or five years and individual seeds may

619

germinate even after being stored for ten years. The seeds of Scots pine likewise have a high percentage of viability for at least three years following maturation of the cones. In this respect certain species may prove to be of advantage to the forester concerned with the artificial regeneration of the forest. The ideal is a tree that begins fruiting at an early age, bears seeds annually or at short intervals, and produces a large number of seeds that retain their viability for a long time. The predominance of Norway spruce and Scots pine in the cultivated forests of Europe is not only because of the value of their wood but also because of their seed production.

Comparison of the seed production of two dominant trees of European forests:

	Norway Spruce *(Picea abies)*	Common Beech *(Fagus sylvatica)*
Age at commencement of seed production	60 years	80 years
Intervals of fruiting	4–5 years	5–10 years
Millions of seeds per hectare	10	0.5
Percentage of viable seeds	70 to 80	60 to 80
Period of viability	4–5 years	6 months

620

620 Rooting coniferous tree cuttings in a greenhouse.

621 Arolla Pine *(Pinus cembra)* seed orchards.

621

622

623

622 A fruiting grafted pine.

623 A felled oak in a flood plain forest.

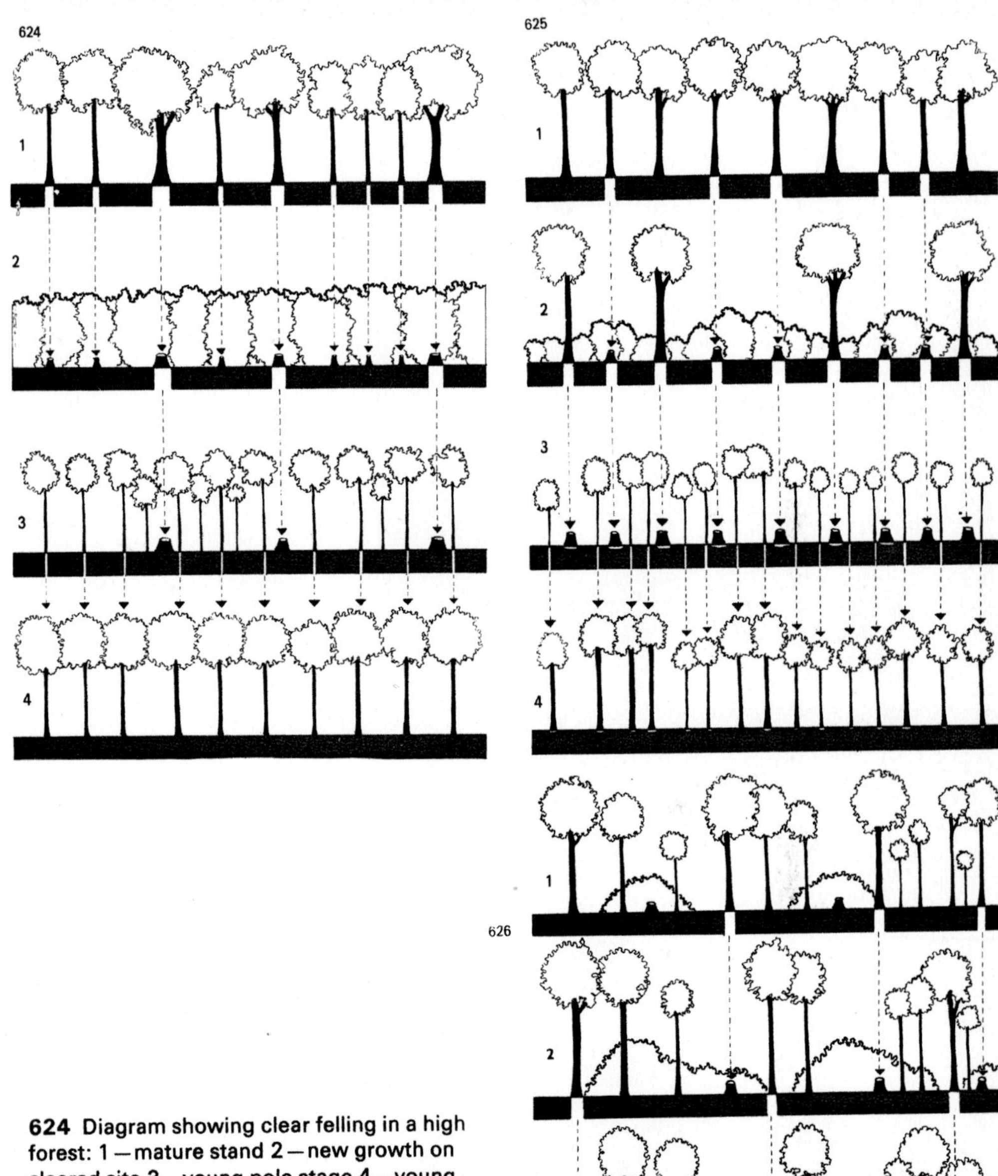

624 Diagram showing clear felling in a high forest: 1—mature stand 2—new growth on cleared site 3—young pole stage 4—young high forest.

625 Diagram showing shelterwood felling in a high forest: 1—mature stand 2—felled single trees and regeneration in gaps 3—pole stage of unequal age 4—high forest of unequal age.

626 Diagram showing selection felling in a high forest; four stages of forest with alternating generations of trees in man-made clearings.

627

628

In former days the collection, extraction and storage of seeds was the task of each individual forester. Nowadays this enterprise is being taken over by specialized firms and large seed establishments where qualified experts supervise the extraction of seeds and their protection against pests, and perform reliable tests of seed viability. What is most important, though, is that they supervise the collection and sale of the seeds of forest trees according to strict geographical and ecological criteria. The forester who carries out artificial regeneration in a given region can be certain that the seed he gets was produced under corresponding climatic and soil conditions.

Rarely is seed sown directly in forests or in large clearings nowadays, except in the artificial spread of certain pioneer trees, for example the introduction of birches to clearings. In this case the forester has at his disposal a great number of readily available seeds and the success of sowing depends in great measure on the condition of the soil and on the weather. Foresters have learned from observing nature that the best method of sowing birch seeds is in the first thin layer of snow.

Seed that is hard to come by and which is acquired at great cost is best grown under the controlled conditions of the forest nursery (Fig. 617 and 619). Here it is possible to prepare the right combination of soil, regulate the microclimate by shading, and provide the necessary moisture by watering. This ensures not only a high percentage of germinating seedlings but also the good growth of the seedlings after they have sprouted. In the nursery all

629

630

627 Canadian logging combine.

628 Swedish logging combine.

629 Lopping with a one-man power saw.

630 Cutting up a fallen tree after a gale.

631

631 Transport of wood using a general-purpose tractor.

632 Skidding wood with a special wheel tractor.

work can be mechanized and the lifting and transport of plants to the forest for purposes of artificial regeneration can also be efficiently organized. For a time it was believed that the optimum conditions provided by the nursery and the damage to roots caused by lifting and transplanting the seedlings was detrimental to the vigour and health of the future trees. This, however, has not proved to be the case in most instances and modern nursery practice even makes use of greenhouses where it is possible to regulate temperature, moisture, light and food with relative precision (Fig. 618). To ease the shock of the move from nursery to clearings, seedlings are hardened off during the period before they are lifted and transplanted.

632

633

Foresters are seeking further ways of improving the quality of forest trees from the genetic viewpoint. They are interested primarily in obtaining plants from top quality trees which measurements have shown to be above average as regards rate of growth, shape of bole and quality of wood. One of the ways to achieve this is to propagate forest trees by means of rooted cuttings from selected specimens (Fig. 620).

Substances promoting the growth of roots on the cut surface of twigs (auxins), as well as better methods of preparing the soil and microclimate, make it possible to propagate the more important forest trees by vegetative means. Equally promising are the results obtained by grafting (Fig. 622). Grafted trees may bear flowers and fruits very quickly so the desired seed is soon available for the production of parent plants. Improved strains obtained by breeding can then be further multiplied by means of seed orchards (Fig. 621).

During the past hundred years regions where forestry has been developed have carried out both unplanned and scientific provenance trials comparing the growth of stands of the same species but from seed of diverse origin. All the trials prove how important are the hereditary traits concealed within the seeds. Within a single species the trees may have a differently arranged root system, differently shaped bole, differently spreading crown, different period of starting growth, and different growth in height and girth. Rarely can important differences be perceived instantly according to the shape of the organs or the overall habit. Only comparative measurements and long-term observations can prove

633 Hauling stacked wood from a forest on peat soil.

634 Logs at the runout at the foot of a mountain slope.

634

635

636

the existence of various climatic races or ecotypes in the case of many important forest trees.

Best known in this respect is the Scots pine, which was the subject of international provenance trials as early as 1907 and many times since. Trees between the ages of twenty and thirty showed marked differences in height. Stands from seeds of the best provenance were twice as tall as those from the worst. The poorest stands came from seed produced in regions having a markedly different climate. For example, on provenance trial areas in German forests the pines that grew best were of Brandenburg and Belgian origin, whereas stands from French and east Russian seed had the lowest height and diameter increment.

However, there were marked differences in other respects just as important to the forester as the volume increment. For a straight bole and low-placed crown, outstanding were pines from seed from east Russia and the Baltic countries. Most resistant to pine needle cast were trees of Finnish origin. All low-quality variants flowered and produced cones much sooner than native and good quality variants.

Of great interest were the results of provenance trials with Norway spruce in Austria, Switzerland and Saxony using the same seed

635 Hauling wood with a hydraulic-arm tractor.

636 Transport of treetrunks by an extraction cableway.

637 Moving valuable logs by a helicopter.

638 Log dump on a mountain pasture.

637

638

639

639 Log dump on a lakeshore in Finland

sown at the same time. These trials were intended to show the differences in growth of trees from different altitudes. The results brought convincing proof that at low altitudes high mountain spruces always grow more slowly than lowland spruces. At median altitudes lowland spruces are still faster growers than high mountain spruces. On trial areas in the subalpine belt spruces of high mountain origin had about the same rate of growth as spruces of lowland origin to begin with. After the twentieth year, however, the scales tipped clearly in favour of the high mountain variants, due largely to the decisive influence of frost damage and snowbreaks. Mountain variants usually started growth earlier in the lowlands and likewise ended growth sooner, whereas lowland variants continued to grow longer in the autumn in mountain and alpine environments, with the result that the new year's growth was damaged by frost.

Similar comparative trials were carried out with various European broadleaved trees (oaks, maples and hornbeams). Extensive provenance trials were also carried out in North American forests with conifers and broadleaved trees. Even though these trials yielded many unexpected or surprising results trees from regions most like the trial area in climate,

640 Loading wood for transport to a paper mill.

641 Mass transport of wood on a lake in Finland.

640

641

642

642 Hauling logs from the forest to an industrial plant.

643 Timberyard of a large forest establishment.

643

Transferring precious tropical timber from railway wagons to ocean-going vessels; Ghana.

Transferring precious tropical timber at the port of Takoradi (Ghana).

Collecting resin from the trunks of Scots Pine *(Pinus sylvestris)*; South Bohemia, Czechoslovakia.

644

soil and geographic location, always had the upper hand. Long-term evaluation has shown that structural differences are always less conclusive evidence than physiological differences, such as rate of growth, length of growth period, hardiness and resistance to parasites.

Foresters will therefore never be content with seeds of unknown provenance. They need to know the place of origin as well as the environment of the seed.

Logging

Even in developed countries where attempts are made to develop multipurpose forests, logging continues to be the most important aspect of forestry (Fig. 623). It is the principal means whereby the forester influences the structure of the forest and determines all further regeneration and tending operations. European foresters distinguish three types of cultivated forest according to the extent of logging: 1, high forest; 2, coppice forest of underwood; and 3, composite forest. IIn a high forest it is the tall exploitable trees (100 to 120 years old) that are felled and regeneration is carried out by sowing seed or planting a new generation of trees in their stead. In a coppice forest only some, mostly younger, trees are felled and regeneration is by vegetative means—by new growth from stumps,

644 Timberyard with barking equipment.

645

646

647

roots, trunk or branches. Composite forest is a temporary form consisting of coppice and standard trees.

Coppiced woodlands are common in the Mediterranean region and at the foot of mountain ranges in the temperate zone, where merchantable species are found which are capable of producing growth from the stump. The cutting cycle in a coppice forest is approximately twenty years and the girth of the felled trunks is correspondingly small. It is best suited for fuel or timber of inferior quality, or for special purposes such as a source of bark for the extraction of tannins (oak coppices), twigs for wickerwork (willow coppices), or poles for vineyards (many acacia coppices). Coppices are common also in tropical regions, mostly round towns where they serve as a source of fuel, for example *Azadirachta indica* or *Cassia siamea* plantations in tropical Africa.

Throughout the world, the prevailing type of cultivated forest is the high forest. In the tropics and temperate regions high forest is formed and maintained by three kinds of felling—clear felling, shelterwood felling and selection felling. In the first instance all the trees on a particular area are felled and the new generation is provided with full sunlight. In shelterwood felling more light is provided for the regenerating trees by felling some of the overwood and opening the canopy in a regular pattern. In selection felling tall exploitable trees are felled here and there so that more light is let in for the young generating trees. Contrary to the opinion of uninformed critics, clear felling was not the outcome of man's desire for total exploitation, but was introduced as a progressive reform designed to prevent disorganized local felling and the ensuing haphazard and chance regeneration of the forest (Fig. 624). This took place sometime in the seventeenth century in Germany where clear felling followed by immediate artificial regener-

645 Spraying beech logs in a timberyard.

646 Timberyard in a forest.

647 Crane-track in front of a modern sawmill in a tropical forest.

648

649

648 Loading huge crates for the overseas transport of machines.

649 Valuable tropical timber at a sawmill in Ghana.

650 Loading timber onto an ocean-going vessel at the port of Turku, Finland.

650

ation was intended to save the totally disorganized and unproductive forests. It was clear felling that made it possible to achieve order in cultivated forests. Clear felling inevitably led to the establishment of spruce and pine monocultures. Problems with the degradation of soil and insect degradations in coniferous monocultures casts a shadow over the clear cutting system. However, equally serious problems brought about by high costs, lack of available work forces and the need for increased mechanization have resulted in the reinstatement of the system in European and North American forests.

In tropical rain forests clear felling is out of the question. The character of the soil, the extremes of the microclimate in clearings and the aggressiveness of forest weeds make shelterwood or selection felling the only possible means of maintaining a cultivated commercial forest. All attempts at clear felling destroy the substance of the forest environment far more rapidly than in temperate regions.

Shelterwood felling was formerly used in Europe primarily in beechwoods and beech-fir woods (Fig. 625). Gradual opening of the canopy depended on the seed year intervals and on the successful establishment of new growth. The combination of natural factors and the forester's activity often met with failure and thus it was necessary to resort to artificial regeneration, which, alas, too often favoured the spruce. In tropical evergreen and semideciduous forests local versions of the shelterwood system are found where the spread of lianas and dwarf trees is vigorously suppressed. In tropical forests shelterwood felling is used with success in south-east Asia in lowland forests which contain mostly trees of the Dipterocarpaceae family.

651 Newly planted stand of the west African Oil Palm *(Elaeis guineensis)*, a forest species with fruits which have many uses.

652 A fenced preserve where wild game is reared.

Everything appears to point to the fact that selection felling comes closest to the natural processes continually taking place in forest undisturbed by man (Fig. 626). Although this system has many supporters and its advantages have been proved many times over, it is not widely used in cultivated forests. The cutting of selected merchantable trees in tropical forests, of course, is not a form of selection felling, for that is a case of individual exploitation with no consideration for the ensuing regeneration of the same species or other quality trees. Selection felling was successfully employed mainly in the beechwood and firwood belt of Switzerland and Germany, though in practice it was often a case of some transitional form, or a combination of selection felling with shelterwood and clear felling. These transition forms are sometimes referred to as group selection cutting. They create small gaps instead of large clearings and these are then enlarged in a direction which provides the best light and moisture conditions for new growth.

651

Because of the dependence of foresters on the labour of farmers and the lack of land in tropical regions, logging and forest regeneration continue to be accompanied by various methods of forest farming. A demarcated area of forest is clear felled, young trees are planted at the desired spacing and farmers are allowed to plant food crops in between. They then tend the young trees while cultivating their own crops for a period of about three years, after which the forester takes responsibility for the trees and organizes felling that favours merchantable trees. In the tropics this is known as the taungya system. The word taungya is derived from a Burmese word, meaning hill cultivation, although the system most probably had its origins in Europe.

Modern forestry techniques

In Chapter 2 mention was made of the problems man has in felling and hauling forest trees. The axe and handsaw were for a long time the only tools the logger had to work with. In the tropics with such tools, a single giant tree with a trunk divided into distinct buttresses could take a whole week to fell. Nevertheless, long before the introduction of efficient power saws, logging remained undaunted by the giant sequoias of California, the enormous eucalyptus trees of Australia and the huge 'mahoganies' of Africa.

Nowadays the felling, lopping and trimming of trees is generally done with a power saw. The light weight and efficiency of these saws has brought about a complete change in logging technology. Large, thick trunks which once required the combined efforts of a team of loggers are now felled by a single man (Fig. 629 and 630). This technology, however, also has its drawbacks, for long-term exposure to the vibrations and noise of the power saw affects the health of the logger. This may be offset by alternating felling with other activities such as barking and skidding.

In recent years technologists have developed machines for felling and multipurpose mechanisms that fell, lop, trim the bole, load

652

653

653 A Red Deer *(Cervus elaphus)* rubs the velvet from its full-grown antlers.

654

654 A wildlife observation hide in Grizedale Forest, Cumbria, North West England.

655

655 A camping site beside Loch Long in the Argyll Forest Park, West Scotland.

656 Children on an organized forest walk in Ennerdale Forest, Cumbria, North West England.

and haul the trees. The efficiency of these small factories on wheels (Fig. 627 and 628) is without doubt remarkable. Their drawback, however, is self-evident—they may be used only in accessible terrain which is not too mountainous, and for concentrated felling over a larger area, in other words clearcutting.

Felled trees must be hauled from the forest. The weight of the wood, irregular ground surface and standing trees make this a very difficult phase of logging, which is why horses are still used for skidding logs (Fig. 633). Even after World War II. loggers still skidded logs on sleds down Europe's steep mountain slopes into the valleys. Sometimes wooden slides were used for the purpose (Fig. 634).

Nowadays, however, variously constructed tractors are the prime movers of wood (Fig. 631, 632 and 635). Their use in logging has saved much labour but has simultaneously posed new problems. Tractors also require greater space for operation and are profitable mostly in concentrated logging on large clear-

656

657

ings. Thus silviculturists are faced once again with the old problem: Is clearcutting to be allowed? And if so, to what extent? And where?

Present-day technology has solved the problem of the removal of timber from rugged mountain terrain with the construction of log extraction cableways (Fig. 636) and helicopters (Fig. 637). As with other mechanisms the setting up of a portable extraction cableway naturally pays only in the case of concentrated logging, but the transport of logs through the air keeps damage both to the logs and the standing trees to the minimum. Even better is the transport of logs by helicopter, which, though extremely costly, is definitely worthwhile in the case of precious woods.

Timber is skidded from stands to a road which can be negotiated by trucks or to a river down which it can be floated. At these log dumps (Fig. 638 and 639) the logs are generally barked, cut into sections, and sorted according to species and quality. It is possible to use further machines in the sorting process (Fig. 640).

From the log dumps the logs travel by truck (Fig. 642), rail or water (Fig. 641) to depots

657 A peaceful picnic in Brechfa Forest, South Wales.

658 A family has a picnic in Rhondda Forest, South Wales.

and timber yards at large ports, railway stations, and the woodworking industry (Fig. 643). Larger timber yards are faced with the problem of preserving the quality of the wood because it may crack if it dries rapidly. The solution is either regular spraying (Fig. 645) or immersion in a water reservoir (Fig. 646). However, logged wood can never be left lying about for long because fungi and insects that damage wood are an ever-present danger.

A well-organized woodworking industry always has smoothly functioning transport to and from the production buildings (Fig. 647). Products of the sawmill may be cut timber (Fig. 649), veneers, plywood, furniture, chipboard or boxes and crates (Fig. 648). For many countries timber production is the chief source of revenue and they have been exporting it for many years (Fig. 650).

Uses of the forest

Mention has already been made of the fact that the forest provided and continues to provide man with other materials, food and drugs. The bark of forest trees, for example, yields tannins required by the leather industry. The leaves, flowers, fruits, twigs and roots of tropical trees as well as forest herbs form part of the diet of the local peoples and are fed to domestic animals. The people of the tropics have always used the starchy pith, fruit and sap of forest palms. The West African Oil Palm *(Elaeis guineensis)*, for example, yields an oil which is used by food and cosmetic industries throughout the world (Fig. 651). Cocoa is also a forest product for it is made from the seeds of the cacao, a small tree grown in the shade of the tropical forest.

Also noteworthy is the amount of resin, rubber, wax and volatile oils yielded by various forests throughout the world. Plantations of Para rubber trees *(Hevea brasiliensis)* continue to exist as specialized cultures, but their importance has declined since the introduction of synthetic rubber. Gaining in importance, on the other hand, is the collection of resin from the Scots pine. Even stands of poorly growing and crooked trees may become a source of great profit.

Even the cultivated forests of Europe and America harbour wild game (Fig. 653) but there are many places in the world where all game birds and mammals have been extermi-

658

659

659 A car park in Dartmoor Forest, South West England.

660 A crowded camping site on the edge of the forest in Glenmore Forest Park, North Scotland.

nated by ruthless shooting and poaching. Wild game in the forest needs to be tended with the same care as new generations of trees. In some forest regions foresters and gamekeepers have become well informed about the biology of the red deer, white-tailed deer, roe-deer, capercaillie, black grouse and pheasant as well as the wild boar, fox, bear, puma, lynx, birds of prey and ravens. They know about their diet and reproduction as well as their diseases and pests. Sections of forest are set aside as game preserves (Fig. 652) which makes it possible to keep a check on the numbers of wild game and to regulate conditions for hunting.

Recreation in the forest

The benefits of forests include more than their produce. In the late twentieth century it is clear to all sensible people that our modern industrial society must preserve forests as an irreplaceable resource where man can find the rest and recreation so necessary to his physical and mental health (Fig. 661). Concrete-weary people need to get away from the noise and polluted atmosphere of the cities now and then, to walk in the wilderness and breathe deeply of the fresh air. Forests must remain also in order to preserve in their gene banks that multiplicity of plants, animals and micro-organisms about which modern science knows virtually nothing.

These are the reasons for the establishment of forest preserves, national parks, and recreational forests having special statutes determined by the requirements of silviculture and tourism. The first forest reservations in Europe were established as early as the first half of the nineteenth century. In North America large national parks are an established tradition. Yellowstone Park, the first national park in the world, was opened in 1872. Later the network of national parks spread throughout the world and they are now to be found far in the north of Scandinavia as well as on the opposite side of the globe, in Tasmania.

Great Britain in 1919 was the last in the league to establish state forests. It is thus as yet without the forestry tradition found in the rest

660

661

661 An important function of the forest is as a place of recreation for city dwellers; forest park in Japan.

A picnic ground in the Everglades National Park; Florida, U.S.A.

Orientation plan and panorama, South West National Park, Tasmania.

662

663

662 Timber-built cabins beside a lake in Kernow Forest, Cornwall, South West England.

663 Holiday forest cabins in among the trees.

664 An open site for forest cabins beside Loch Lubnaig in Strathyre Forest, Perthshire, Central Scotland.

665 A view point from a forest walk in Coed-y-Brenin Forest, North Wales.

of Europe, as well as being the most densely populated country in that continent, with most of its people concentrated in huge urban areas. Pressure from these millions of town dwellers, increasingly mobile in their family cars, for recreational facilities in the countryside has resulted in parts of the comparatively young forests being designated 'forest parks'—seven in all throughout the country (Fig. 659). Within these forest parks, and in other large areas of woodland, many kinds of activities are encouraged, both formal and incidental. Some people prefer the solitude which the forest gives and for them the nature trails, the marked walks, the fishing in the forest lakes and streams, or just strolling among the trees and sitting on the benches set up at view points along the forest trails, are the pleasures sought. Others like their recreation to be more sociable and enjoy pony trekking through forest rides and clearings or boating on the lochs and lakes, and picnicking in the special sites provided (Fig. 657 and 658). Collective activities for students and youth clubs such as orienteering and environmental studies are also very popular (Fig. 656).

Although most people make only day trips to the forests, a great many spend their annual holidays there. Caravan and camping sites within the forest bring all the facilities within

664

665

walking distance of the holiday base, and for those who want something more sophisticated among the trees, there are timber-built forest cabins to be hired, containing all the basic domestic needs for a family (Fig. 655, 660, 662, 663 and 664). Here, in the heart of the forest, escape from the town is complete; bird song at sunrise, glimpses of deer as dusk creeps over the trees and easy access to the nature observation hides, the arboreta, and the nature trails with their on-the-spot descriptions of animals, birds and plants, as well as to all the other activities which the varied interests of a family group may require (Fig. 654 and 665).

In these ways city folk, individually, in families or in larger groups, are encouraged to visit the forests, and from these visits comes an increasing appreciation of the vital role which forests play in the life of a nation. For, without timber and its derivatives few of man's activities can prosper, and without facilities for outdoor recreation in the countryside very many urban dwellers would find life infinitely more drab. The forest provides these assets to protection, for within it thousands of people can be absorbed in a comparatively small space without feeling crowded, and while they are there they can share and enjoy nature's most advanced and varied habitat with the many thousands of plants and creatures which make it so

Though national parks and reservations cover large areas they are unable to embrace the great diversity of forest ecosystems originally formed on the earth. Besides, nowadays not even the largest reservations are protected against the global pollution that is spreading from urban and industrial centres. The safe future of forests can be assured only by international conventions on the environment and by the rational mobilization of all economic and scientific forces in defence of the remaining centres of forest flora and fauna.

It remains to be hoped that man will not sever the limb he sits on, that human civilization will not disrupt the biome on which it is so dependent, and that forests will not become a wonder, shown to the public only on the payment of a high entrance fee.

666

666 A mixed mountain forest is important to man in many different ways.

INDEX